Max Planck

WHERE IS SCIENCE GOING?

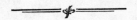

MAX PLANCK

MAX PLANCK

WHERE IS SCIENCE GOING?

— WITH A PREFACE BY —

ALBERT EINSTEIN

Translated and Edited
by
JAMES MURPHY

OX BOW PRESS
WOODDBRIDGE, CONNECTICUT
1981

First Published in English January 1933

1981 Reprint published by
Ox Bow Press
P.O. Box 4045
Woodbridge, Connecticut 06525

Reprinted by arrangement with
George Allen & Unwin Ltd.

ISBN 0-918024-21-8 (Hardcover)
ISBN 0-918024-22-6 (Paperback)

Library of Congress Card Number 80-84974

Printed in the United States of America

CONTENTS

PREFACE

BY

ALBERT EINSTEIN

MANY kinds of men devote themselves to Science, and not all for the sake of Science herself. There are some who come into her temple because it offers them the opportunity to display their particular talents. To this class of men science is a kind of sport in the practice of which they exult, just as an athlete exults in the exercise of his muscular prowess. There is another class of men who come into the temple to make an offering of their brain pulp in the hope of securing a profitable return. These men are scientists only by the chance of some circumstance which offered itself when making a choice of career. If the attending circumstance had been different they might have become politicians or captains of business. Should an angel of God descend and drive from the Temple of Science all those who belong to the categories I have mentioned, I fear the temple would be nearly emptied. But a few worshippers would still remain—some from former times and some from ours. To these latter belongs our Planck. And that is why we love him.

I am quite aware that this clearance would mean the driving away of many worthy people who have built a great portion, and even perhaps the greatest portion, of the Temple of Science. But at the same time it is

obvious that if the men who have devoted themselves
to science consisted only of the two categories I have
mentioned, the edifice could never have grown to its
present proud dimensions, no more than a forest could
grow if it consisted only of creepers.

But let us forget them. *Non ragionam di lor.* And let
us fix our gaze on those who have found favour with
the angel. For the most part they are strange, taciturn
and lonely fellows. Yet, in spite of this mutual resem-
blance, they are far less like one another than those
whom our hypothetical angel has expelled.

What has led them to devote their lives to the
pursuit of science? That question is difficult to
answer and could never be answered in a simple
categorical way. Personally I am inclined to agree with
Schopenhauer in thinking that one of the strongest
motives that lead people to give their lives to art and
science is the urge to flee from everyday life, with its
drab and deadly dullness, and thus to unshackle the
chains of one's own transient desires, which supplant
one another in an interminable succession so long as
the mind is fixed on the horizon of daily environment.

But to this negative motive a positive one must be
added. Human nature always has tried to form for
itself a simple and synoptic image of the surrounding
world. In doing this it tries to construct a picture
which will give some sort of tangible expression to
what the human mind sees in nature. That is what
the poet does, and the painter, and the speculative
philosopher and the natural philosopher, each in his
own way. Within this picture he places the centre of

gravity of his own soul, so that he will find in it the
rest and equilibrium which he cannot find within the
narrow circle of his restless personal reactions to
everyday life.

Among the various pictures of the world which are
formed by the artist and the philosopher and the
poet, what place does the world-picture of the theoretical physicist occupy? Its chief quality must be a
scrupulous correctness and internal logical coherence,
which only the language of mathematics can express.
On the other hand, the physicist has to be severe and
self-denying in regard to the material he uses. He has
to be content with reproducing the most simple
processes that are open to our sensory experience,
because the more complex processes cannot be represented by the human mind with the subtle exactness
and logical sequence which are indispensable for the
theoretical physicist.

Even at the expense of completeness, we have to
secure purity, clarity and accurate correspondence
between the representation and the thing represented.
When one realizes how small a part of nature can thus
be comprehended and expressed in an exact formulation, while all that is subtle and complex has to be
excluded, it is only natural to ask what sort of attraction this work can have? Does the result of such
self-denying selection deserve the high-sounding name
of World-Picture?

I think it does; because the most general laws on
which the thought-structure of theoretical physics is
built have to be taken into consideration in studying

even the simplest events in nature. If they were fully known one ought to be able to deduce from them by means of purely abstract reasoning the theory of every process of nature, including that of life itself. I mean *theoretically*, because in practice such a process of deduction is entirely beyond the capacity of human reasoning. Therefore the fact that in science we have to be content with an incomplete picture of the physical universe is not due to the nature of the universe itself but rather to us.

Thus the supreme task of the physicist is the discovery of the most general elementary laws from which the world-picture can be deduced logically. But there is no logical way to the discovery of these elemental laws. There is only the way of intuition, which is helped by a feeling for the order lying behind the appearance, and this *Einfuehlung* is developed by experience. Can one therefore say that any system of physics might be equally valid and possible? Theoretically there is nothing illogical in that idea. But the history of scientific development has shown that of all thinkable theoretical structures a single one has at each stage of advance proved superior to all the others.

It is obvious to every experienced researcher that the theoretical system of physics is dependent upon and controlled by the world of sense-perception, though there is no logical way whereby we can proceed from sensory perception to the principles that underlie the theoretical structure. Moreover, the conceptual synthesis which is a transcript of the empirical world may be reduced to a few fundamental laws on which the whole synthesis is logically built. In every important advance the physicist finds that the fundamental laws are simplified more and more as experimental research advances. He is astonished to notice how sublime order emerges from what appeared to be chaos. And this cannot be traced back to the workings of his own mind but is due to a quality that is inherent in the world of perception. Leibniz well expressed this quality by calling it a pre-established harmony.

Physicists sometimes reproach the philosophers who busy themselves with theories of knowledge, claiming that the latter do not appreciate this fact fully. And I think that this was at the basis of the controversy waged a few years ago between Ernst Mach and Max Planck. The latter probably felt that Mach did not fully appreciate the physicist's longing for perception of this pre-established harmony. This longing has been the inexhaustible source of that patience and persistence with which we have seen Planck devoting himself to the most ordinary questions arising in connection with physical science, when he might have been tempted into other ways which led to more attractive results.

I have often heard that his colleagues are in the habit of tracing this attitude to his extraordinary personal gifts of energy and discipline. I believe they are wrong. The state of mind which furnishes the driving power here resembles that of the devotee or the lover. The long-sustained effort is not inspired by any set plan or purpose. Its inspiration arises from a hunger of the soul.

I am sure Max Planck would laugh at my childish way of poking around with the lantern of Diogenes. Well! why should I tell of his greatness? It needs no paltry confirmation of mine. His work has given one of the most powerful of all impulses to the progress of science. His ideas will be effective as long as physical science lasts. And I hope that the example which his personal life affords will not be less effective with later generations of scientists.

INTRODUCTION

MAX PLANCK
A BIOGRAPHICAL SKETCH

BY

JAMES MURPHY

ONE day in June 1932 I paid a visit to Albert Einstein at his summer home in Caputh, some fifteen miles west of Berlin. We had a long-drawn-out tea together on a multitude of topics, from the chances of the various political parties at the coming election to the chances of somebody finally discovering a simple formula for the unification of all physical laws. The house is pitched high on a terraced slope and overlooks a beautiful lake. Level with the upper story there is a veranda which is like the spacious platform of an observatory station. And there is a telescope with which Einstein amuses himself by gazing on the stars. When dusk came on, and the blazing sunlight that had been beating on the lake all day was turning to a mellow glow, we went for a stroll on the veranda to watch the sunset and while away the time until the evening meal would be ready. Within doors the political crisis had been the central topic of conversation; but here, amid the natural harmony of lake and forest and sinking sun, a higher theme made its appeal.

The name of Max Planck came into our talk, and

the various philosophical problems which quantum physics have given rise to. To my more sweeping generalities Einstein would most invariably reply *"Nein, das kann man nicht sagen."* But when I put forward something more qualified he would reflect for a while and say *"Ja, das können Sie sagen."* We were agreed, I think, that though the relativity theory has captured the imagination of the world, the quantum theory has been a more fundamental force in bringing about the modern revolution in scientific thought.

While we were on this point I asked Einstein to write me an introduction for a book of essays by Planck, to be published in English. Einstein shied at the suggestion. He said that it would be presumptuous on his part to introduce Max Planck to the public; for the discoverer of the quantum theory did not need the reflected light of any lesser luminary to show him off. That was Einstein's attitude towards Planck, expressed with genuine and naïve emphasis.

I explained that the book in question would be for the general public and that, though the name of Planck is a household word in Germany and with scientists all the world over, he is not so popular in English-speaking countries as the founder of the relativity theory. Einstein did not consider this a very regrettable circumstance. He would have been pleased if the truth were the other way round. But my point was that it is a good rule of logic to define the less known through the better known, no matter what the objective merits of the one or the other may be. He submitted to the force of this argument and agreed to a short introduction but insisted that it must be short, for anything long would be pretentious.

The present chapter is not an enlargement on Einstein's introduction. It is meant rather to be a biographical sketch of a purely objective kind. My first task here is to indicate the place which the author of the following chapters holds in the modern development of physical science. Then I shall endeavour to describe for the reader, as simply and as vividly as I can, the personality of Max Planck—his scie fic career, his attitude towards the function of theoretical physics as an intellectual force in the modern everyday world, his philosophy of life, his contemporary activities as a citizen and man of learning, and finally, his place and prestige among his own people.

The first part of this task will be best discharged if I leave it to a few leaders amongst Planck's colleagues to define the place he holds in the general picture of modern scientific progress.

What significance has the name of Max Planck in the history of Physics? The answer to that question can be indicated by pointing to the position which a portrait of Max Planck would occupy in a pictorial representation illustrating the development of science. At the end of a long gallery there is a turning and a wide space or angle of the wall. On that space the portrait of Max Planck hangs, with one hand taking grateful leave of the classical past and the other pointing to a new corridor where the paint is hardly yet dry on the portraits that hang there—Einstein, Niels Bohr, Rutherford, Dirac, Eddington, Jeans,

B

Millikan, Wilson, Compton, Heisenberg, Schroed-inger, etc., etc. Sir James Jeans, in his popular little book *The Mysterious Universe*, describes the position thus:[1]

"At the end of the nineteenth century it first became possible to study the behaviour of single molecules, atoms and electrons. The century had lasted just long enough for science to discover that certain phenomena, radiation and gravitation in particular, defied all attempts at a purely mechanical explanation. While philosophers were still debating whether a machine could be constructed to reproduce the thoughts of Newton, the emotions of Bach or the inspiration of Michelangelo, the average man of science was rapidly becoming convinced that no machine could be constructed to reproduce the light of a candle or the fall of an apple. Then, in the closing months of the century, Professor Max Planck of Berlin brought forward a tentative explanation of certain phenomena of radiation which had so far completely defied interpretation. Not only was his explanation non-mechanical in its nature; it seemed impossible to connect it up with any mechanical line of thought. Largely for this reason, it was criticized, attacked and even ridiculed. But it proved brilliantly successful, and ultimately developed into the modern 'quantum theory,' which forms one of the dominating principles of modern physics. Also, although this was not apparent at the time, it marked the end of the mechanical age in science, and the opening of a new era."

[1] *The Mysterious Universe*, 1932 edition, pp. 16 and 17.

Another British scientist, Lord Rutherford, gives the following estimate of his German colleague:

"The name of Planck is a household word among the scientific men of all countries and all unite in their admiration for his great and enduring contributions to Physical Science.

It is difficult to realize to-day, when the quantum theory is successfully applied in so many fields of science, how strange and almost fantastic this new conception of radiation appeared to many scientific men thirty years ago. It was difficult at first to obtain any convincing proof of the correctness of the theory and the deductions that followed from it. In this connection I may refer to experiments made by Professor Geiger and myself in 1908. On my side, the agreement with Planck's deduction of e (e is the elementary electric charge and the value is expressed in electrostatic units) made me an early adherent to the general idea of a quantum of action. I was in consequence able to view with equanimity and even to encourage Professor Bohr's bold application of the quantum theory propounded by Planck."[1]

The significance of Planck's achievement is thus described by Niels Bohr, the famous Danish physicist:

"Scarcely any other discovery in the history of science has produced such extraordinary results within the short span of our generation as those which have directly arisen from Max Planck's discovery of the elementary quantum of action. This discovery has been prolific, to a constantly increasing degree of

[1] *Die Naturwissenschaften*, Vol. 26, p. 483.

progression, in furnishing means for the interpretation and harmonizing of results obtained from the study of atomic phenomena, which is a study that has made marvellous progress within the past thirty years. But the quantum theory has done something more. It has brought about a radical revolution in the scientific interpretation of natural phenomena. This revolution is a direct development of theories and concepts which originated from the pioneering work done by Max Planck in studying cavity radiation. Within the past thirty years these theories and concepts have grown and expanded into that scientific elaboration which is called quantum physics. The picture of the universe formed on the lines of quantum physics must be looked upon as a generalization that is independent of classical physics, with which it compares favourably for its beauty of conception and the inner harmony of its logic.

I should like emphatically to call attention to the consequences of this new knowledge. It has shattered the foundations of our ideas not only in the realm of classical science but also in our everyday ways of thinking. It is to this emancipation from inherited traditions of thought that we owe the wonderful progress which has been made in our knowledge of natural phenomena during the past generation. That progress has gone beyond even the highest hopes to which it gave rise a few years ago. And the present state of physical science can probably be indicated best by saying that nearly all the lines of thought which have led to fruitful results in experimental research have naturally blended together into a common harmony without thereby losing their individual fertility. For having placed in our hands the means of bringing about these results the discoverer of the quantum theory deserves the unqualified gratitude of his colleagues!" [1]

One name more will be sufficient to add to this distinguished list. It is that of Professor Heisenberg, the Leipzig physicist, who is the founder of the now popular Theory of Indeterminacy. Heisenberg writes as follows:

"In 1900 Max Planck published the following statement: *Radiant heat is not a continuous flow and indefinitely divisible. It must be defined as a discontinuous mass made up of units all of which are similar to one another.*

At that time he could scarcely have foreseen that within a span of less than thirty years this theory, which flatly contradicted the principles of physics hitherto known, would have developed into a doctrine of atomic structure which, for its scientific comprehensiveness and mathematical simplicity, is not a whit inferior to the classical scheme of theoretical physics."[2]

Let us come now to the personal story of Max Planck himself. He was born at Kiel, Germany, on April 23, 1858. His father was Professor of Constitutional Law at the University and was afterwards transferred to Goettingen in the same capacity. The chief work whereby his name is known is the Prussian Civil Code,

[1] *Die Naturwissenschaften*, Vol. 26, p. 483.   [2] *Ibid.*, p. 490.

of which he is co-author. It is often said that the great physicist has inherited certain qualities from his father, especially the juridical faculty of sifting experimental evidence, disentangling the significant from the meaningless and probing to the absolute values hidden beneath the relative. He has also a faculty for constructive clarity in building up a mathematical synthesis. But perhaps the most striking quality which he has derived from his early family associations is shown in his attitude towards physical science as a branch of human culture, forming an integral part with the other branches of human learning and exercising its influence on the destiny of humanity not merely in a material way but even more deeply in a spiritual way.

When Max Planck was seventeen years old he entered the University of Munich, taking physics as his chief subject. Three years later he went to Berlin to complete his course at the University there. At that time Helmholtz and Kirchhoff were the leading scientific lights of the Prussian Capital. Kirchhoff was Professor of Physics at the University and young Planck read under him there, also attending the lectures of Helmholtz and Weierstrass. He always asserts that Kirchhoff was responsible for his keen interest in thermodynamics, especially the famous Second Law. It was on this subject that Max Planck wrote his treatise for the doctorate, which he presented at the University of Munich a year later, in 1879, when he received the doctorate *Summa cum Laude*. The treatise was entitled *De secunda lege fundamentale doctrinae mechanicae caloris.* Perhaps I ought to explain here that in qualifying for the taking of degrees all universities in Germany are treated as one. A student may take part of his course in one university and part in another; so that, in case he should wish to follow some special line of work in which there is an eminent professor in some university away from his home town, he can attend there and indeed make the rounds of all the eminent professors if he likes, from one university to another. The sum-total will be credited to him as if he had studied at the one university all along.

Having received his doctorate, Max Planck became a *Privat Dozent* at Munich University. The *Privat Dozent* is a university lecturer who receives fees but no salary. In 1885 Planck was appointed Professor of Physics at the University of Kiel and in 1889 he came to Berlin as Professor Extraordinarius there. In 1892 he was appointed full professor in succession to Kirchhoff at the University of Berlin. In 1912 he became Permanent Secretary to the Prussian Academy of Science. In 1919 he received the Nobel Prize for Physics. And in 1926 he became Professor Emeritus, Schroedinger succeeding him in the Berlin Chair of Theoretical Physics. In 1930 Adolf Harnack died and Max Planck was elected President of the Emperor William Society for the Advancement of Science, which is the highest academic post in Germany.

What was it that first put Planck on the trail of the quantum? That would be a rather long story to tell; for the telling of it would involve an account of the various attempts that were being energetically made

towards the end of the last century to solve the spectroscopic riddle of heat radiation. As this expression
may not convey a very clear idea to the mind of the
average reader, it will be well to explain it a little.

Everybody is acquainted with the solar spectrum,
which results in the breaking up of white light by
passing it through a prism, thus producing a spectrum
of coloured rays which group themselves on the screen
and run continuously from red to violet. Newton was
the first to handle the phenomenon in a scientific
manner, and this led to the great problem of the nature
of light itself. In the case of heat radiation we have
a corresponding phenomenon. Sir William Herschel
was the first to show that the solar spectrum is not
confined to that part which is visible to the eye, from
the red to the violet. In 1800 he discovered that there
are infra-red solar rays. By applying a thermometer to
the successive colours he discovered an uneven distribution of heat in the solar spectrum, the heat being
greatest below the red. This inequality had never
previously been suspected.

Now it is a matter of everyday experience that a
body when moderately heated gives out an invisible
radiation. The frequency of the undulations is too low
to influence the eye. As the temperature is gradually
increased, in a piece of iron for instance, one might
expect that violet rays would first be perceptible, as
these have the minimum wave-length which is necessary
to stimulate the sense of sight. But that is not what
happens. The light is at first dull red, then bright red,
and finally becomes glowing white. Now the question

here is, how does the intensity of the rays of different frequency change with the rising temperature? This is what is called the problem of the spectral distribution of radiation for different temperatures. It is the problem to which Max Planck devoted the first twenty years of his academic career. In his address before the Royal Swedish Academy of Science in Stockholm, on the occasion of receiving the Nobel Prize, he said:

"Looking back over the past twenty years to the time when the idea of the physical quantum of action, and the measurement of it, first emerged into definite shape from a mass of experimental facts, and looking back beyond that over the long and labyrinthine path which finally led to the discovery, I am vividly reminded of Goethe's saying that men will always be making mistakes as long as they are striving after something. During such a long and difficult struggle the researcher might be tempted again and again to abandon his efforts as vain and fruitless, except that every now and then a light strikes across his path which furnishes him with irrefutable proof that, after all his mistakes in taking one by-path after another, he has at least made one step forward towards the discovery of the truth that he is seeking. The steadfast pursuance of one aim and purpose is indispensable to the researcher and that aim will always light his way, even though sometimes it may be dimmed by initial failures.

The aim which I had for so long before my mind was the solution of the distribution of energy in the normal spectrum of radiant heat. Gustav Kirchhoff had shown that the nature of heat radiation is completely independent of the character of the radiating bodies. This pointed to the existence of a universal function which must be dependent exclusively on temperature and wave-length but in no way dependent on the properties of the substance in question. If this remarkable function could be discovered it might give a deeper understanding of the relationship between energy and temperature, which forms the main problem of thermodynamics and consequently of molecular physics as a whole. At that time no way suggested itself of discovering this function except to select from the various bodies in nature certain kinds of bodies whose capacities for emitting and absorbing heat are known and then to calculate the heat radiation when the exchange of temperature is stationary. According to Kirchhoff's theory, this must be independent of the nature of the body itself."

He then traced in a modest and objective way the rocky road that he had followed, the slips and falls by the wayside, the discouragement, but always the persistent effort and the determination to win through. Finally the goal was reached, after a long journey of twenty years.

Planck first presented the results of his discovery in a communication to the German Physical Society, on December 14, 1900. His paper was entitled "On the Distribution of Energy in a Normal Spectrum." The discovery of the function mentioned above had been arrived at in the shape of a formula for measuring radiant energy. He had experimented with what is known as cavity radiation. This means that he heated a hollow body to incandescence and allowed a beam of radiation to issue through a small opening and analysed the beam in the spectroscope. In this way it was found that radiant energy is not a continuous *flow*. It is emitted in integral quantities, or *quanta*, which can be expressed in integral numbers. In other words, the measurement always results in integral multiples of hv, where v is the frequency and h is a universal constant, now known as Planck's constant. His greatest triumph of technical skill was in deducing the value of this constant to be $6 \cdot 55 \times 10^{-27}$ erg-seconds. No radiation can be emitted unless it is of at least that amount or an integral multiple thereof. That is to say, our stove cannot give us any heat until it has accumulated at least that amount. Then it will not increase the radiation of its heat until it accumulates another integral packet which is exactly double that amount, and so on. We can have $2\,hv$ and $3\,hv$ and $4\,hv$; but we cannot have any fractional parts of $h\,v$. This involved a revolutionary concept for radiation of heat, and the concept was eventually shown to extend to all radiation and finally to the interior structure of the atom itself.

It soon became evident that Planck had brought to light something that not merely explained the puzzle of the spectrum of radiant heat but something that is universally fundamental in nature. This was shown by the gradual application of his theory in all directions. Within a few years after its promulgation Einstein applied the quantum theory to explain the constitution of light and showed that light follows the same process as heat radiation and is emitted in parcels or quanta, called photons. Physicists in every country began to practise the technique of "Quantizing" and achieved very remarkable results. H. A. Lorentz, the famous Dutch scientist, put the case thus in 1925.

"We have now advanced so far that this constant (Planck's universal h) not only furnishes the basis for explaining the intensity of radiation and the wavelength for which it represents a maximum, but also for interpreting the quantitative relations existing in several other cases among the many physical quantities it determines. I shall mention only a few; namely, the specific heat of solids, the photo-chemical effects of light, the orbits of electrons in the atom, the wavelengths of the lines of the spectrum, the frequency of the Roentgen rays which are produced by the impact of electrons of given velocity, the velocity with which gas molecules can rotate, and also the distances between the particles which make up a crystal. It is no exaggeration to say that in our picture of nature nowadays it is the quantum conditions that hold matter together and prevent it from completely losing its energy by radiation. It is convincingly clear that we are here dealing with real relations because the values of h as derived from the different phenomena always agree, and these values differ only by slight shades from the number which Planck computed twenty-five years ago on the experimental data that were then available".[1]

[1] *Die Naturwissenschaften*, Vol. 35, 1925, p. 1008.

It is not the place here to attempt an explanation of the scientific aspects of the quantum theory. The reader will find several popular accounts—some of them perhaps all too popular—of Planck's revolutionary theory in various books on modern science. My task here is rather to indicate the source from which the material of this book has originated and try to explain why it is that Planck has felt the need to assert himself so strongly in dealing with certain philosophical aspects of contemporary science. Most of the essays here—the discussion on positivism and the discussion on determinism and free will—are outside the sphere of pure physics. Why is it that the *doyen* of German physicists has felt himself called upon to take so strong a stand?

A great deal has been written about the philosophical implications of the quantum theory. Some of the physicists declare categorically that the development of the quantum theory has led to the overthrow of the principle of causation as an axiom in scientific research. Sir James Jeans puts this side of the question as follows:

"Einstein showed in 1917 that the theory founded by Planck appeared, at first sight at least, to entail consequences far more revolutionary than mere discontinuity. It appeared to dethrone the law of causation from the position it had heretofore held as guiding the course of the natural world. The old science had confidently proclaimed that nature could follow only one road, the road which was mapped out from the beginning of time to its end by the continuous chain of cause and effect; state A was inevitably succeeded by state B. So far the new science has only been able to say that state A may be followed by state B or C or D or by innumerable other states. It can, it is true, say that B is more likely than C, C than D, and so on; it can even specify the relative probabilities of states B, C and D. But, just because it has to speak in terms of probabilities, it cannot predict with certainty which state will follow which; this is a matter which lies on the knees of the gods—whatever gods there be."[1]

Further on Sir James Jeans states:

"Or again, to take another analogy, it is almost as though the joints of the universe had somehow worked loose, as though its mechanism had developed a certain amount of 'play,' such as we find in a well-worn engine. Yet the analogy is misleading if it suggests that the universe is in any way worn out or imperfect. In an old or worn engine, the degree of 'play' or 'loose jointedness' varies from point to point; in the natural world it is measured by the mysterious quantity known as 'Planck's constant h,' which proves to be absolutely uniform throughout the universe. Its value, both in the laboratory and in the stars, can be measured in innumerable ways, and always proves to be precisely the same. Yet the fact that 'loose jointedness,' of any type whatever, pervades the whole universe *destroys the case for absolutely strict causation*, this latter being the characteristic of perfectly fitting machinery."[2]

[1] *The Mysterious Universe*, 1932, pp. 17 and 18.   [2] *Ibid.*, p. 24.

The italics are mine. Sir James Jeans's assertion is illustrative of an attitude that is fairly common among modern physicists. But it is an attitude to which Planck is stoutly opposed. Scientifically considered, it is premature; and, logically considered, it is too much of a jump towards a sweeping conclusion. Planck would claim, and so would Einstein, that it is not the principle of causation itself which has broken down in modern physics, but rather the traditional formulation of it. The principle of causation is one thing; but the way in which it was formulated by Aristotle and the Scholastics and Newton and Kant is quite another thing. As applied to happenings in nature, whether in the sphere of mind or of matter, the traditional formulation must be considered rather too rough-and-ready. In the discussion appended to this book the latter point will be examined somewhat more sharply. What is of chief interest here is to ask why Planck considers the causal controversy of so much importance that he spends a considerable portion of his time to-day—and he is a very busy man—in the delivery of lectures and the writing of essays on it. Why does he assert himself so emphatically on this point? The answer cannot be that he is a stickler for the authority of tradition; because, as a matter of fact, he has headed the greatest revolt in modern science. The answer therefore must be looked for in a different direction.

At the present time there is a wave of public interest in physical science. It arose immediately after the war and shows no signs of receding. This is undoubtedly due to the fact that physical science is the most vital expression of the higher activities of human thought to-day. Moreover the metaphysical content of the higher speculations in theoretical physics seems to be the favourite modern pabulum for the soul-hunger which was formerly appeased by the ideals of art and religion. From many points of view, this may be a fortunate thing; but from other points of view it may be a misfortune, especially from the scientific point of view. Edwin Schroedinger has recently published a brilliant essay (*Ist die Natur-wissenschaft Milieubedingt?* Barth, Leipzig, 1932) in which he suggests that physical science has fallen a victim to the *Zeitgeist*. To-day the *Umsturzbedürfnis* (The need for something radically different from the established order) is a universal feature of our civilization. The authority of tradition is a drawback rather than a recommendation in the case of principles or methods hitherto dominant in art or music or even politics and business. And we find this same devaluation influencing scientific ideas. When Einstein promulgated his relativity theory much of the enthusiasm with which it was proclaimed was associated with the impression that it constituted a complete overthrow of Newtonian doctrines; whereas, as a matter of fact, relativity is an expansion and refinement of Newtonian physics. And so when Heisenberg proclaimed his Principle of Indeterminacy it was almost immediately interpreted, even among physicists themselves, as definitely effecting an overthrow of the causation principle. As a matter of fact, we have no means whatsoever of proving or disproving the existence of

causation in the external world of nature. And the
aim which Heisenberg had before his mind in formulating the Principle of Indeterminacy was to find a rule
whereby we can deal with minute processes in natural
phenomena, such as those in which the elementary
quantum of action is involved. Here the causal
principle is not applicable. That is to say, we cannot
estimate simultaneously both the velocity and position
in time-space of a particle and say where it will be a
moment hence. But this does not mean that the causal
sequence is not actually verified objectively. It means
that we cannot detect its operation; because, as things
stand to-day, our research instruments and our mental
equipment are not adequate to the task. The Principle
of Indeterminacy is in reality an alternative working
hypothesis which takes the place of the strictly causal
method in quantum physics. But Heinsenberg himself
would be one of the first to protest against the idea
of interpreting his Principle of Indeterminacy as
tantamount to a denial of the principle of causation.

Why is it then that this hasty conclusion is so much
in vogue? It is probably due to two things: First the
Zeitgeist. The spirit of the age does not want to be
considered the heir of the old order and wishes to
consider itself free from all laws handed down through
the authority of tradition. Secondly, the standardization
of modern life, with its mass production and high-
powered salesmanship and advertisement and transport
and mass housing and insurance undertakings, etc.,
has evolved a system of statistical rules which are true
when masses of events are concerned, though they

are not at all applicable to the individual. People call this the principle of statistical causality. Physicists have brought it over into their science and often speak of it as the opposite of strict causation in the classical sense. They speak of statistical causation as opposed to dynamical causation. But, as a matter of fact, statistical causality and even what are called the laws of probability are all based on the presupposition of strict causation in the individual cases dealt with. According to the statistical causality principle of insurance companies, so many thousand people die of certain diseases in the year, at certain ages and in certain professions. It is on the basis of these statistics that insurance policies are drawn up. But these statistics have nothing to do with the actual cause of death in the case of the individual insured.

Now, anybody who has the interests of his own art or science at heart will strive to protect it against adulteration through the intrusion of principles and methods which are foreign to it. That is exactly Planck's position in regard to physical science. If we are living at a time that is breaking away from the old political and social traditions, this is fundamentally because the old traditions are not suitable to the changed economic, and therefore social, order in which we live. But scientific research is something that has to be carried on apart from the changing circumstances of human existence. It is natural, of course, that the public mind should turn to that branch of our spiritual culture which is the most vital to-day, namely physical science, and seek in it a *point d'appui* for the general world-outlook. But this very fact alone, flattering as it may be for the individual scientist, endangers the integrity of the science in question.

It is from this source that Planck's interest in the causality controversy arises. And it is in this light too that we are to view his attitude towards the positivist thesis. The undue popularization of physical science has probably tempted some physicists hastily to build up a theoretical structure wherein the public mind may find a congenial object of awe and wonderment and, in a sense, worship, such as in former days was supplied by the mysteries of religion. This may explain that phase of modern theoretical science which somewhat resembles the sophist phase into which Greek philosophy degenerated and which also characterized the decadence of the scholastic movement. It was this latter decadence that instigated the founding of the empiricist school in England at the time of Locke, for the purpose of reconstructing a reliable basis of philosophic thought. We have a similar movement in physical science to-day, with a similar purpose in view. There are some physicists who would reduce the scope of physical science to a bald description of the events scientifically discovered as occurring in nature, and would entirely exclude all theory and hypothesis-building. Planck feels that this restriction of scope is anti-scientific and very much to the detriment of physics. That is why he is so stoutly opposed to it. As the *doyen* of international physicists he feels within his rights in taking up the cudgels against the renunciatory movement. That he voices the mind of leading German scientists in this regard I am quite convinced. Not long ago I happened to be at dinner with a number of Planck's colleagues at Goettingen. Hermann Weyl was there, and Max Born, and James Franck. Planck was mentioned a great deal and there was some rather lively discussion about his intransigence on the causation principle, but one and all agreed in championing his stand against the positivist teaching.

As this is a sort of close-up sketch, for the purpose of bringing the personality of the author of the quantum theory vividly before the mind of the reader, I shall conclude with a few remarks on Planck's personal standing among his colleagues. He is undoubtedly the most popular figure in the academic world of Germany. Indeed one may say without the slightest fear of exaggeration that he is the beloved of his colleagues. Professor Sommerfeld of Munich, whose name is also renowned in the realm of quantum physics, wrote of Planck some time ago: "His doctor's diploma (in 1879) bore the superscription *Summa Cum Laude*. We would write the same superscription over his work during the whole fifty years that have elapsed since then, and not for his scientific work alone but also for his human example. He has never written a word that was not genuine. And in polemical questions he has always been chivalrous to his opponent. When the German Physical Society was being reorganized there was dissension and antagonism; but Planck was the trusted representative of both sides, the naturally fair-minded arbiter."

Sommerfeld tells a story about Planck that is illustrative of the unselfish and modest manner in which
he is always ready to collaborate with his colleagues.
Sommerfeld was once engaged on some research
concerning what is known as Phase-space in atomic
physics. He wrote to Planck for assistance, and Planck
immediately placed at his disposal the results of
his own experiments in the same field. Sommerfeld
fell into a poetical strain and sent a couplet to Planck
in which he explained that he himself was only putting
forth a humble endeavour to gather a few flowers in
the great new land of quantum physics which Planck
had turned from an unknown wilderness into arable
ground.

> *Der sorgsam urbar macht das neue Land*
> *Dieweil ich hier und da ein Blumenstraeuschen fand.*

To this delightful compliment Planck replied with a
quatrain in a still more gentle spirit.

> *Was Du gepflueckt, was ich gepflueckt*
> *Das wollen wir verbinden,*
> *Und weil sich eins zum andern schickt*
> *Den schoensten Kranz draus winden.*

> (What you have picked and I have picked,
> These we shall bind together.
> Entwining thus a fair bouquet
> From gifts we send each other.)

In the modest little account which Planck gave of
himself before the Royal Swedish Academy on the
occasion of receiving the Nobel Prize, he mentioned
a tragedy which has afflicted his family life. This was

the loss of his two daughters, both of whom died soon
after marriage, one might almost say in their bridal
robes, and the loss of a very gifted son in the war.
Another son was wounded but has survived and is
now a Minister in the von Papen Cabinet.

When conversing with Planck even on scientific
subjects one often feels that this tragedy of his children
has made a deep impress on his soul. The memory of
it seems to evoke a certain wistful quality which is
profound in his nature and gives it the warmer glow
that one is inclined to call mystic. And indeed, though
a scientist and a perfectly practical man of the world
and an up-to-date gentleman in manner and dress and
also a sportsman, who climbed the Jungfrau to cele-
brate his seventy-second birthday a few years ago—
still one often thinks of him in conjunction with
Beethoven, I don't know why, and one remembers
that at the beginning of Planck's career there was a
question whether he would develop the musical side
of his genius or the scientific side. He developed the
latter. But he could not develop the one without
enriching the other also, because the pursuit of
theoretical science demands as its first prerequisite
the constructive imagination of the artist. And the
constant seeking after nature's harmonies responds to
the longing for musical expression. Anyhow it is a
significant fact that the two greatest scientists in
Germany, Einstein and Planck, are musicians also.

On visiting his home in the Wangenheimer Strasse,
Berlin, and chatting with him in that big room which
is at once his reception parlour and study, I often

think that his own private trials have been sublimated
by the tragedy of his country, and this in its turn
sublimated by the universal tragedy of the modern
world. For on this he broods more than most busy
men do. But the moment the first cloud of depression
shows itself he counters it with his favourite motto
Man muss optimist sein. We must be optimists. He has
said that the inscription on the gate of the temple of
science indicating the condition on which alone her
devotees may enter, is: *Ye must have Faith*. Running
through all his work and all that he has said or says
there is always this golden thread of a living faith in
the ultimate purposes of creation.

WHERE IS SCIENCE GOING?

CHAPTER I

FIFTY YEARS OF SCIENCE

HERE I shall give a short sketch of physical science in Germany during the period of my own active work in that field. For the sake of clarity it will be better if we ignore the chronological sequence of events and try to trace the main lines along which the various specific groups of ideas have developed. While doing this I shall take into account also the co-operative work done by scientists in other countries. And if I mention certain names, while leaving out many others quite as eminent if not more so, these names will be cited merely as landmarks to indicate a particular stage or turning-point, without any suggestion whatsoever of making a personal valuation of the work done by the scientist mentioned.

Let us take the year 1880 as our starting-point. At that time four great names shone out above all others to illuminate the direction along which physical inquiry was advancing. These were: Hermann von Helmholtz, Gustav Kirchoff, Rudolf Clausius and Ludwig Boltzmann. The two former were the chief luminaries in the contiguous provinces of mechanics and electrodynamics, while the two latter were prominent in the associated spheres of thermodynamics and atomic physics. But there was really no dividing space between the activities of those four pioneers. They represented a concept of the physical universe which was common to them all and towards which their attitudes were in the closest harmony. That common concept rested on a twofold foundation. One part of the foundation consisted of Hamilton's Principle of Least Action, which includes the Principle of Conservation of Energy. The second part of the foundation represented the Second Law of Thermodynamics.

At that time it was considered by all physicists as practically certain that any subsequent development in theoretical physics must necessarily be in the direction of working out those two universal principles to their final conclusion and application. Nobody then dreamt that within a short period of time the two principles which stood so proudly alone in supporting the structure of physical science would have to take other principles into partnership on an independent and equal footing.

The advent of these new principles was already foreshadowed in some of the ideas put forward by the older pioneers whom I have mentioned and also in the tendencies of those who then represented the rising generation. Heinrich Hertz was the most outstanding among the latter. He stands at the opening of the new era and it would be impossible to overestimate his services to the cause of modern physics. Unfortunately his work was cut short by an early death, at the age of thirty-four, while he was still active as Professor of Theoretical Physics at the University of Bonn. Despite his epoch-making discovery of the propagation of electromagnetic waves through a vacuum, Hertz was not the founder of a new scientific doctrine. What he achieved was to bring an already existing theory to its completion, for he finally established the Maxwell theory of light and thus displaced all the various other theories which for a long time had been struggling against one another for precedence in the field of electrodynamics. By reason of these achievements Hertz must be credited with the accomplishment of a very important advance towards the unification of theoretical physics, for he thus brought optics and electrodynamics under the one doctrinal discipline.

His last work was the simplification of Newtonian mechanics to an ideal degree. In Newtonian mechanics the distinction had always been drawn between kinetic and potential energy as essentially different entities. Hertz succeeded in unifying this bipartite concept, which he did by fundamentally eliminating the idea of a force. The Newtonian *force* was identified by Hertz with internal motion in matter, so that what had hitherto been called potential energy was now replaced by the kinetic concept. Hertz however never attempted to explain the nature of these inner motions in any particular direction, such as gravitation for instance. He contented himself with establishing in principle the hypothesis of unification.

If we make allowance for certain theories that were still only in what may be called a rudimentary stage of development, we can say that at the end of last century the science of theoretical physics as a whole presented the imposing aspect of a complete and perfectly articulated structure. A penetrating observer however could not have failed to notice that in some sections of the foundation there were open flaws which could not be looked upon with anything like satisfaction. Hertz did not fail to see this. And he did not fail to call attention to the fact that the integration of the structure here would prove at least very difficult if not impossible. These flaws soon became the object of attack on the part of scientific criticism. And this criticism developed into a creative movement which eventually brought about the most important expansion that theoretical physics has experienced since the time of Newton.

No doctrinal system in physical science, or indeed perhaps in any science, will alter its content of its own accord. Here we always need the pressure of outer circumstances. Indeed the more intelligible and comprehensive a theoretical system is the more obstinately it will resist all attempts at reconstruction or expansion. And this is because in a synthesis of thought where there is an all-round logical coherence any alteration in one part of the structure is bound to upset other parts also. For instance, the main difficulty about the acceptance of the relativity theory was not merely a question of its objective merits but rather the question of how far it would upset the Newtonian structure of theoretical dynamics. The fact is that no alteration in a well-built synthesis of thought can be effected unless strong pressure is brought to bear upon it from outside. This strong pressure must come from a well-constructed body of theory which has been firmly consolidated by the test of experimental research. It is only thus that we can bring about the surrender of theoretical dogmas hitherto universally accepted as correct. And thus only can we succeed in forcing a fundamental revision of the whole doctrinal structure. Following a reconstruction of this type there invariably arises a fresh series of problems for experimental research to tackle. It is in the tackling of these problems that new ideas are suggested which subsequently lead to the formulations of further theories and hypotheses.

This alternative play of theory and experiment, of theoretical constructions on the side of abstract reason and the testing of these by their application to objective reality, is a special characteristic of modern physics. Indeed it is of enormous significance in all scientific progress, for it is the one safe and sound source from which reliable and enduring results can be produced.

There were two problems of theoretical physics which may be said to have absorbed almost the whole of Hertz's attention towards the end of his life. But they defied all his attempts at solution. And these two problems eventually became the nucleus out of which the physics of our day have developed. These problems were: (1) the nature of the cathode rays, and (2) electrodynamic motion. Each of these two problems has its own history; for each furnished the starting-point of an independent development. The former led to the theory of electrons, the latter to the relativity theory.

THE ELECTRONIC THEORY

The cathode rays were first discovered by von Pleucker in the year 1859. This discovery naturally opened the question as to the nature of the rays themselves. Were they the carriers of electric charges or were they undulatory, like the rays of light? The fact that the X-rays could not be deflected by bringing a magnet to bear on them seemed to point to their electrical character. But Hertz decided in favour of the opposite view. He came to this conclusion after numerous experiments in which he had tested the cathode rays by bringing them to bear on a magnetic needle and found in each case that the needle remained in its position of equilibrium. Hertz accordingly was inclined to identify the cathode rays with the waves of light-ether, which scientists had for a long time been vainly trying to discover. If Hertz were right here his theory would mean that one of the awkward voids in the structure of theoretical physics would thus be filled in.

But, contrary to Hertz's suggestion, there were indications which pointed to the assumption that the cathode rays are corpuscular and the carriers of electric charges. With the advance of experimental methods scientists began to believe more and more that the cathode rays would eventually be found to be the carriers of negative electricity. Indications pointed definitively in this direction once W. Wien had discovered the electric charge in the rays and D. Wiechert their velocity. Therewith the foundation of the electronic theory was led.

It is interesting to note how in this case theory and experiment worked hand in hand, one taking the leadership to-day and the other to-morrow. Experiment first appears in the lead, represented especially by Phillipe Lenard. In 1892 he showed that the cathode rays could pass through very thin metal foils and he succeeded in obtaining them outside the tube in which they were generated. Later on the experimental impulse produced a marvellous and unexpected result, in 1895, when W. Roentgen, while working on the cathode rays, discovered the X-rays and thus with one blow opened up a new kingdom for physical science. At the same time his discovery placed a completely new task before the theoretical physics of the time. This led indirectly to the discovery of uranium rays on the part of the French physicist, Henri Becquerel. A further development in the same experimental field eventuated in the discovery of the radioactive substances and the establishment of the theory of radioactivity, on the part of Rutherford and Soddy.

Experimental investigation into the nature of the various phenomena connected with cathode rays and X-rays and radioactivity progressed on all sides. The special problem to be solved was that of their origin and the nature of their activity. But the Roentgen rays for a long time absolutely defied all attempts at

quantitative analysis. In the early experimental stages
it was readily established that the X-rays were of an
electromagnetic nature, through the fact that when
we put a piece of metal opposite the cathode inside
the tube—the so-called anti-cathode—streams of electrons are shot off from the anti-cathode. Yet it was
for a long time impossible to arrive at any satisfactory
results in measuring the wave-length of the X-rays.
Here it was that the work of a theorist, Professor
von Laue, opened the way for the next decisive
step.

In the year 1912 von Laue, in collaboration with the
experimental physicists, W. Friedrich and P. Kipping,
succeeded in ascertaining the wave-length of the
X-rays by passing them through crystalline media
and thus bringing about the phenomena of interference. In this way it was found possible to measure
the wave-length, but the experiment holds good of
course only for homogeneous Roentgen rays, because
otherwise confusion would arise from the various
interference positions overlapping one another.

Von Laue's discovery turned out quite as valuable
in the sphere of atomic physics as in the sphere of
optics. It enabled physicists definitely to classify the
Roentgen rays and the Gamma rays with the radioactive substances in electrodynamics. On the other
hand, the carriers of the cathode rays—that is to say,
the free electrons—with their relatively small mass,
proved to be something entirely new to physical
science. It was the introduction of these electrons that
made it possible to understand various physical pheno-

mena which hitherto had remained in the region of mystery.

As far back as 1881 Helmholtz pointed out, in his famous Faraday lecture, that from the standpoint of chemical atomics the empirically deduced laws of chemical decomposition by galvanic action could be explained only in case we attribute an atomic structure to electricity as well as to matter. The atom of electricity postulated by Helmholtz first appeared in the cathode rays, free and detached from all matter, and was again located in the Beta rays of radioactive substances. In contradistinction to chemical atoms, all electric atoms are found to be uniform and to differ from one another only in their velocity. The discovery of electrons and the introduction of them into the scientific picture of the universe threw a new light on the nature of metallic conduction. It is well known that an electric current when passing through a metal conductor, such as an ordinary piece of copper wire, produces no chemical change. Once the existence of electrons became known it seemed natural to consider these free electrons as carriers of the electric current through the metal. This opinion, which had previously been put forward by Wm. Weber, was now revived and further developed by E. Riecke and P. Drude.

Once the free electrons had been accepted by physical science as veritable factors in nature, an attempt was made to prove that these electrons also existed in a "bound" condition. This attempt put the investigators on the track of a whole new series of physical and chemical properties of matter. P. Drude

D

explained the optical dispersion and chemical valency of a substance by referring these to the electrons in the atoms and for this purpose he differentiated between firmly bound and loosely bound electrons. The former cause dispersion of light and the latter account for the property of chemical valency. Subsequently H. A. Lorentz formulated the whole electronic theory as a complete and independent synthesis. His special endeavour was to ascertain if and how far all material constants of a substance can be accounted for by the arrangement and interaction of the atoms and electrons contained in them.

Taking the results thus obtained, together with the work done in the sphere of radioactivity, the final consequence of the researches which were directed towards discovering the inner constitution of matter within the past fifty years is the knowledge that all matter is made up of two primordial elements: negative electricity and positive electricity. Both consist of uniform minute particles containing uniform but opposite charges. The positive particle, which is the heavier, is called the proton and the negative, the lighter, is called the electron. The union of both is called the neutron. Every electrically neutral chemical atom is made up by a certain number of protons held fast together and by an equal number of electrons of which some are bound to the proton and form together with it the nucleus of the atom, while the others—that is to say, the free electrons—move in orbits around the nucleus. The number of these latter, which are called free or orbital electrons, gives in each case what is called the atomic number. It is on this number that all the chemical properties of the various elements depends.

THE RELATIVITY THEORY

I have spoken at length of Heinrich Hertz and his work in the movement which eventually led to the establishment of the electronic theory. Now we come to the second great theory which I have mentioned as forming, with the electronic theory, one of the twin principles which were entirely undreamt of fifty years ago and are now among the main piles supporting the scientific structure. This second principle is the relativity theory. And here again we find that Hertz was among the pioneers. The last and most fruitful period of his life's work was devoted largely to the study of electrodynamic phenomena in moving bodies. In this work Hertz chose as his starting-point the principle that all movement is relative. Adopting Maxwell's theory as his groundwork, he formulated for the phenomena of electrodynamic movement a system of equations in which the velocity of the bodies concerned is taken only in a relative sense. This is expressed by the fact that the equations, just as the Newtonian laws of motion, remain unchanged if the velocity of the body in question be taken in relation to a moving reference system or, in other words, a moving observer. There is no necessity in the Hertzian theory to introduce the idea of a special substantial medium of transmission for the electrodynamic waves. If we should think of introducing ether as a substantial medium of transmission here, then we must assume that it has no independent motion of its own in relation to matter but is completely carried forward with it.

The Hertzian theory was excellent in its inner coherence, but from the beginning he recognized that it had considerable drawbacks. A wave of light passing through air which is also in motion must be considered in conjunction with the movement of the air, just as in the case of sound waves, no matter how rarefied the air may be. This was a necessity of the Hertzian theory but it was contradicted through a decisive discovery made by Fizeau, who proved that light passes through moving air with just the same velocity as through still air. In other words it goes against the wind or in a perfect calm or with the wind at the same rate of speed.

H. Lorentz endeavoured to smooth out this contradiction between the Hertzian theory and Fizeau's discovery by putting forward the idea of a stationary ether permeating the whole of space. This was suggested as the carrier and transmitter of all electrodynamic action. In this ether the atoms and electrons move about as distinct particles. Thus the advantages of Hertz's theory were retained and at the same time the theory could be harmonized with Fizeau's findings. On the other hand, however, this involved the renunciation of the idea of relativity; because it established a definite object of reference absolutely at rest. This was a static ether and the hypothesis of its existence seemed more satisfactory than any that had hitherto been put forward.

The relativity principle in this way received a set-back. But reprisals were soon forthcoming, inasmuch as new defects arose which the Hertzian theory could not cope with. All attempts to measure the absolute velocity of the earth had failed. In other words it turned out impossible to measure the velocity of the earth in relation to the hypothetically static ether. Even the most delicate of all experiments, namely that carried out by Michelson and Morley, could detect no trace whatsoever of the influence of the earth's motion on the velocity of light although, according to the Lorentzian doctrine, this should have made itself felt.

Under these circumstances theoretical physics, at the end of last century, was faced with the alternative of renouncing either the remarkably useful Lorentzian theory or the theory of relativity. The crisis came into public notice very strikingly at a meeting of the Society of German Physicists and Physicians,[1] which was held at Dusseldorf in the August of 1898. On that occasion the whole question was discussed in a debate which centred around two papers that were read there, one by W. Wien and the other by H. A. Lorentz. The controversy remained open for seven years. Then, in the year 1905, a solution was put forward by Albert Einstein in his Theory of Relativity. The Einsteinian hypothesis allowed the Lorentzian theory to stand, but only at the cost of introducing what at first sight appeared to be an entirely alien hypothesis, namely, that the dimensions

[1] Gesellschaft deutscher Naturforscher und Aerzte.

of time and space cannot be taken independently of one another but must be welded together when there is question of the velocity of light *in vacuo*. This hypothesis was logically unassailable, because it was expressed in a mathematical formulation that was flawless in itself. Yet the relativity thesis completely contradicted all hitherto accepted opinions.

Only a few years after Einstein had published his first presentation of the relativity hypothesis Minkowski succeeded in bringing a corroborative light to bear on the suggestion. He showed that if we look upon time as something imaginary and assume the unit of time to be the amount of time which a beam of light takes to travel over the unit of length, then all our electro-dynamic equations in relation to space and time will be symmetrical; because the one dimension for time and the three dimensions for space enter as factors on an equal footing in the formulation of every law of electrodynamics. Thus the three-dimensional "space" is expanded into the four-dimensional "world" and the mathematical laws that govern the whole field of electrodynamics remain invariable when the reference system—that is to say, the observer—changes its velocity, just as they remain invariable when the reference system changes its motion from one direction to another.

Now the next question to arise was this: If the relativity hypothesis in its new formulation is to have meaning and validity for physical science as a whole, it must apply not merely to electrodynamics but also to mechanics. If however the relativity theory be applicable and valid in the field of mechanics then we must change the laws of motion formulated by Newton; because the Newtonian laws do not remain constant when the four-dimensional reference system is changed. Out of these problems arose what is called relativist mechanics, which are an expansion and refinement of Newtonian mechanics. The theory of relativist mechanics was verified by experiment in the case of rapidly moving electrons, for this experiment showed that mass is not independent of velocity. In other words it was shown that the mass of a rapidly moving body increases with the increase of velocity. And thus a further corroboration of the Einstein hypothesis was provided.

Beyond the achievement of welding space and time together with the mechanical laws of motion, the relativity theory accomplished another and no less important amalgamation. This was the identification of mass with energy. The unification of these two concepts establishes for all equations in physical science the same kind of symmetry as the four co-ordinates of the space-time continuum, the momentum vector corresponding to the place vector and the energy scalar corresponding to the time scalar. Another important consequence of the relativity theory is that the energy of a body at rest is given a quite positive value, which is expressed through the multiplication of its mass by the square of the velocity of light; so that in general mass is to be considered under the concept of energy.

But Einstein did not rest content with this success

of his theory. Once it had been shown that all reference systems, or standpoints of observation, are equally valid as long as they are interchanged with one another through linear rectangular transformation, Einstein was led to ask whether and how far such an equivalence would hold good for a quite arbitrary reference system. The transformation of simple mechanical equations to any other reference system generally involves certain additional factors such as that of a centrifugal force where there is question of a rotating reference system, such as the earth, and these additional factors appear as the effect of gravity in so far as ponderable mass is identified with inertial mass. Now the hypothesis that, from the viewpoint of physical science, no geometrical reference system has from the outset any advantage over any other system, and that the property of invariance can be explained only on the basis of the Riemann fundamental tensor—which on its part depends on the distribution of matter in space—led to the formulation of the general relativity theory. This general theory of relativity includes the former theory as a special case and holds the same relation to the special theory of relativity as Riemann's geometry holds to Euclid's geometry.

The practical significance of the general theory of relativity is naturally confined to very powerful gravitational fields such as that of the sun, whereby the colour and the light are affected, or to movements which have secular periods, such as the perihelion displacement of the orbit of Mercury. The general theory of relativity represents the first great step towards the ideal goal of geometrizing the whole of physics. Einstein has recently devoted himself to the task of opening the way for the second step, which would unite mechanics and electrodynamics under the one system of equations. To this end he has undertaken the task of formulating a single field theory, based on a different geometry from that of Riemann. We have yet to await the final success of this attempt.

THE QUANTUM THEORY

Apart from and quite independent of the relativity theory, the quantum theory has given a new impress to theoretical physics during the past thirty years. Just as in the case of the relativity theory, its origin and foundation arose from recognition of the fact that the old classical theory had to be abandoned because it failed to explain results which had been established through experimental means. These results, however, were not obtained in the region of optics but rather in that of thermodynamics and arose from the measurement of radiant energy in the emission spectrum of black bodies.

According to the Kirchhoff law this radiant energy is independent of the nature of the radiating substance and therefore has a universal significance. In this direction indeed the classical theory had already led to important results. In the first place L. Boltzmann deduced from Maxwell's discoveries, in regard to the pressure exerted by radiation, and from the laws of thermodynamics, the dependence of all types of radiation on temperature. W. Wien extended the same principle further and showed that the curve of the distribution of energy in the spectrum, especially in its location and maximum extent, is displaced by a change of temperature. This was in full harmony with the most delicate measurements. But in relation to the shape of this curve there resulted a very strong discrepancy between the conclusions arrived at theoretically and the measurements carried out by von Lummer and Pringsheim, Rubens and Kurlbaum. Then Max Planck, taking the laws of thermodynamics as the basis upon which an explanation of the experimental results could be obtained, arrived at the revolutionary hypothesis that the manifold features which an oscillating and radiating picture possesses are complete entities in themselves and that the difference between any two features of the picture is characterized by a definite universal constant, namely the elementary quantum of action.

The establishment of this hypothesis involved a fundamental break with the opinions hitherto held in physical science; because until then it had been an accepted dogma that the state of a physical picture could be indefinitely altered. The fruitfulness of the new hypothesis showed itself immediately in the fact that it led to a law which explained the distribution of energy in the spectrum and was in perfect harmony with the measurements. But it also supplied a means for determining the absolute weights of molecules and atoms. Up to then, in so far as atomic realities had been measured at all, science had to be content with more or less rough estimates. Einstein readily showed that the new theory had a further consequence inasmuch as it applied to the energy and specific heat of material bodies. Hitherto it had been only a mere supposition that specific heat decreases illimitably with decrease of temperature, but this now became established by experimental proof. Max Born and Th. von Karman, on the one hand, and P. Debye, on the other, began to study carefully from the standpoint of the quantum theory the problem of the dependence of specific heat on temperature, and succeeded in formulating a law according to which it is possible to reckon the variation of specific heat with temperature from the elastic constants of the substance in question. The most striking proof, however, for the universality of the quantum of action is to be found in the circumstance that not only the whole of the heat theory put forward by W. Nernst in the year 1906, independently of the quantum theory, is in harmony with the quantum theory, but also that the chemical constant introduced by Nernst depends on the quantum of action. This was clearly demonstrated by O. Sackur and H. Tetrode.

Belief in the soundness of the quantum theory has nowadays become so strong and widespread that if the measurement of a chemical constant does not tally with the theoretical reckoning the discrepancy is attributed, not to the quantum theory as such, but to the manner of its application, namely the assumption of certain atomic conditions in regard to the substance in question. But the laws of thermodynamics are only of a summary and statistical nature and can give only summary results when applied to electronic processes in the atom. Now if the quantum of action has the significance which has come to be ascribed to it to-day in thermodynamics it must make itself felt also in every single process within the atom, in every case of emission and absorption of radiation and in the free dispersion of light radiation. Here it was Einstein once again who formulated the hypothesis that the light quanta have an independent existence and exercise an independent activity.

This led to the putting forward of a whole series of new questions and started correspondingly new investigations in physics and chemistry. These dealt with the emission of light quanta on the one hand and with electrons, atoms, and molecules, on the other. The first direct measurement of the quantum of action was obtained by J. Franck and G. Hertz by liberating quantities of light through electronic impulses. Niels Bohr succeeded in further elucidating the theory and extended its application beyond the thermodynamic sphere. On the basis of the quantum he was able to lay down the laws which are followed by the minute activities taking place in the interior world of the atom. By the construction of his atom model he showed mathematically that if the electrons of the atom be held to revolve at enormous speeds, the change of energy involved in the displacement of an electron from one orbit to another exactly corresponds to the quantum theory that the variation of the physical state does not take place gradually but in integral jumps. This was the first time that the quantum theory

came to be applied outside of the region of thermodynamics.

The quantum way of solving physical problems was
further extended by A. Sommerfeld, who in this
manner succeeded in solving the riddle of delicate
spectral structures which had hitherto defied explanation. Independently entirely of spectral phenomena, the
Bohr model of the atom proved effective in the eluci-
dation of chemical laws, including those underlying
the periodically occurring functions of elements in
chemical structures.

Professor Bohr himself has never claimed that his
model of the atom provides the final solution of the
quantum problem; but the correspondence principle
which he introduced has proved remarkably fruitful
because, in combination with the classical theory, it
points out the direction for the further development
of the quantum theory.

In point of fact a certain amount of uncertainty
lingered on because the discontinuous character of
the Bohr atom, the so-called stationary electron orbits,
did not accord in their peculiarities with the laws of
classical mechanics. Professor Heisenberg discovered
a way out of this difficulty by formulating a detailed
description of electronic motion in a sense entirely
foreign to classical theories. He showed that only
dimensions which in principle were directly measurable
should be treated theoretically, and thus he succeeded in formulating certain equations by which the
problem of applying the quantum theory has been
solved in regard to its universal validity. The close

relation between this particular method of reckoning and that of matrix computation was brought to light by the collaboration of Max Born and P. Jordan, and a further significant step in this direction was accomplished by W. Pauli and P. Dirac.

It is remarkable how such a roundabout way, which even sometimes appeared to run in opposite directions, led to the selfsame goal and opened up new territory which has extended the basis of the quantum theory. A further extension followed, with the founding of the wave-theory. The Heisenberg theory originally recognized only integral magnitudes in the quantities measured. That is to say, his results verified the condition of discontinuity postulated by the quantum theory. But another and complementary interpretation developed independently of Heisenberg, out of suggestions first made by L. de Broglie. The Einstein light quanta are of a twofold nature. Looked at from the viewpoint of energy, they act as discrete and divisible particles—that is to say, they are concentrated quanta, or photons; but if we consider them from the electromagnetic standpoint all experiment has shown that they are like a spherical wave or pulse spreading in all directions, completely corresponding to the Maxwellian wave-theory of light. This is one of the great dilemmas of modern physics. And the hypothesis of wave-mechanics is an attempt to solve it. It was E. Schroedinger who first presented an exact analytical formulation of wave-mechanics, in the partial differential equations adduced by him. For the integral values of energy, on the one hand, this led directly to the quantizing rules which Heisenberg had laid down, while on the other hand it extended the grounds of application of the quantum theory to disintegrating processes and even more tangled problems. At the present stage of its development we can safely say that the theory of wave-mechanics has definitely established itself as a generalization and expansion of the classical corpuscular mechanics. The difference between classical mechanics and wave-mechanics arises principally from the circumstances that the laws of motion in respect of a physical picture cannot be formulated as they were formulated in classical mechanics—that is to say, the picture cannot be broken up into infinitesimally small fractions and the movement of each fraction dealt with independently of the others. On the contrary, according to wave-mechanics, the picture must be held before the eye as a whole and its movement must be looked upon as arising from the individual and mutually differentiated integral movements. From this it follows closely that, not the local force—as in Newtonian mechanics—but the integral force—that is to say, the potential—enters the fundamental equations. Moreover it follows that there can be no sense in talking about the state of a particle in the sense of meaning its position and velocity. This state at best is rather a certain underlying space for the play of dimensional ordering of the quantum of action. Therefore in principle every method of measurement involves an uncertainty in regard to the corresponding sum-total.

It goes without saying that the laws of nature are

in themselves independent of the properties of the instruments with which they are measured. Therefore in every observation of natural phenomena we must remember the principle that the reliability of the measuring apparatus must always play an important rôle. For this reason many researchers in quantum physics are inclined to set aside the principle of causation in the measurement of natural processes and to adopt a statistical method in its place. But instead of this I think it may be suggested with equal justice that we might alter the formulation of the causal principle as we have received it from classical physics, so that it may again have its strict validity. But this question as to the rival merits of the strictly causal and statistical methods will depend upon how far the one proves more fruitful of results than the other.

CHAPTER II

IS THE EXTERNAL WORLD REAL?

WE are living in a very singular moment of history. It is a moment of crisis, in the literal sense of that word. In every branch of our spiritual and material civilization we seem to have arrived at a critical turning-point. This spirit shows itself not only in the actual state of public affairs but also in the general attitude towards fundamental values in personal and social life.

Many people say that these symptoms mark the beginnings of a great renaissance, but there are others who see in them the tidings of a downfall to which our civilization is fatally destined. Formerly it was only religion, especially in its doctrinal and moral systems, that was the object of sceptical attack. Then the iconoclast began to shatter the ideals and principles that had hitherto been accepted in the province of art. Now he has invaded the temple of science. There is scarcely a scientific axiom that is not nowadays denied by somebody. And at the same time almost any nonsensical theory that may be put forward in the name of science would be almost sure to find believers and disciples somewhere or other.

In the midst of this confusion it is natural to ask whether there is any rock of truth left on which we can take our stand and feel sure that it is unassailable and that it will hold firm against the storm of scepticism

E

raging around it. Science, in general, presents us with the spectacle of a marvellous theoretical structure which is one of the proudest achievements of constructive reasoning. The logical coherence of the scientific structure was hitherto the object of unstinted admiration on the part of those who criticized the fundamentals of art and religion. But this logical quality will not avail us now against the sceptics' attack. Logic in its purest form, which is mathematics, only co-ordinates and articulates one truth with another. It gives harmony to the superstructure of science; but it cannot provide the foundation or the building-stones.

Where shall we look for a firm foundation upon which our outlook on nature and the world in general can be scientifically based? The moment this question is asked the mind turns immediately to the most exact of our natural sciences, namely, Physics. But even physical science has not escaped the contagion of this critical moment of history. It is not merely that the claim to reliability put forward by physical science is questioned from the outside; but even within the province of this science itself the spirit of confusion and contradiction has begun to be active. And this spirit is remarkably noticeable in regard to questions that affect the very fundamental problem of how far and in what way the human mind is capable of coming to a knowledge of external reality. To take one instance. Hitherto the principle of causality was universally accepted as an indispensable postulate of scientific research, but now we are told by some physicists that it must be thrown overboard. The fact that such an extraordinary opinion should be expressed in responsible scientific quarters is widely taken to be significant of the all-round unreliability of human knowledge. This indeed is a very serious situation, and for that reason I feel, as a physicist, that I ought to put forward my own views on the situation in which physical science now finds itself. Perhaps what I shall have to say may throw some light on other fields of human activity which the cloud of scepticism has also darkened.

Let us get down to bedrock facts. The beginning of every act of knowing, and therefore the starting-point of every science, must be in our own personal experiences. I am using the word, experience, here in its technical philosophical connotation, namely, our direct sensory perception of outside things. These are the immediate data of the act of knowing. They form the first and most real hook on which we fasten the thought-chain of science; because the material that furnishes, as it were, the building-stones of science is received either directly through our own perception of outer things or indirectly through the information of others, that is to say, from former researchers and teachers and publications and so on. There are no other sources of scientific knowledge. In physical science we have to deal specially and exclusively with that material which is the result of observing natural phenomena through the medium of our senses, with of course the help of measuring instruments such as telescopes, oscillators and so on. The reactions thus registered in observing external nature are collated and schematized on the basis of repeated observations and calculations. This subject-matter of our scientific constructions, being the immediate reactions of what we see, hear, feel, and touch, forms immediate data and indisputable reality. If physical science could discharge its function by merely concatenating these data and reporting them, then nobody could question the reliability of its foundations.

But the problem is: Does this foundation fully meet the needs of physical science? If we may say that it is the business of physical science, solely and exclusively, in the most accurate and most simple way, to describe the order observed in studying various natural phenomena, then is the task of physical science adequately and exhaustively fulfilled? There is a certain school of philosophers and physicists who hold that this and this alone forms the scope of physical science. Many outstanding physicists have been induced to accept this view because of the general confusion and insecurity that arise from the sceptical spirit of the times. They feel that here at any rate is a foundation that is impregnable. The school which puts forward this view is generally called the Positivist School; and in all that I have to say here I shall take the word Positivism in that sense. Since the time of Auguste Comte, the founder of Positivism, many meanings have been given to the word. Therefore I think it well to declare here at the outset that I am restricting its application to the definite meaning which I have already indicated. This happens also to be the meaning in which the word, Positivism, is most generally used.

Now let us ask, is the foundation which Positivism offers broad enough to support the whole structure of physical science? The best test that can be applied in finding an answer to this question is to ask where Positivism would lead if we once were to accept it as offering the sole groundwork of physical science.

Suppose for the moment that we are positivists. And let us take the trouble to control ourselves so that we shall hold strictly to its logical implications and not allow commonplaces and considerations of sentiment to lure us from the logical train of positivist thought. Let us here and now decide that no matter what singular consequences we may encounter in dealing with the positivist line of thought we shall stick steadfastly to it. And we shall be sure that in doing so we cannot be faced with logical contradictions directly emerging from the field of observation; because obviously two actually observed facts in nature cannot be in logical contradiction to one another. On the other hand as long as we remain positivists we must deal with every kind of experience and ignore no source of human knowledge whatsoever. Therein lies the strength of the positivist theory. As long as physical science sticks to the positivist rule it occupies itself with all the problems that can be answered through direct observation. Every problem that has a meaning of definite importance comes within the ambit of physical science under the positivist rule. If we are to content ourselves with a direct observation of natural phenomena and the recording of them, we shall obviously have no fundamental riddles to solve nor any obscure questions. Everything will lie in the open daylight. Thus far the state of affairs looks quite simple. But it is no simple matter at all to carry out the principle when we begin to deal with individual cases. Our daily habits of speech make it rather difficult for us to observe the strict positivist rule. In ordinary life when we speak of an outer object—a table, for instance—we mean something that is different from the table as actually observed by physical science. We can see the table and we can touch it and we can try its firmness by leaning on it and its hardness and if we give it a thump with our knuckles we shall feel hurt. In the light of positivist science the table is nothing more than a complex of these sensory perceptions and we have merely got into the habit of associating them with the word *table*. Remove these sensory perceptions and absolutely nothing remains. In the positivist theory we must entirely ignore everything beyond what is registered by the senses and therefore we are impregnable in this clearly defined realm. For the positivist, to ask what a table in reality is has no meaning whatsoever; and this is so with our other physical concepts. The whole world around us is nothing but an analogue of experiences we have received. To speak of this world as existing independently of these experiences is to make a statement that has no meaning. If a problem dealing with the external world does not admit of being referred immediately to some kind of sensory experience and does not allow of being placed under observation, then it has no meaning and must be ruled out. Therefore within the scope of the positivist system there is no place for any kind of metaphysics. If we glance upwards at the star-strewn firmament we see innumerable points or patines of light which move in a more or less regular way through the heavens. We can measure the intensity and the colour of their rays. According to the positivist theory, these measurements are not merely the raw material of astronomy and astrophysics, but they are the sole and exclusive subject-matter of these sciences. Beyond merely recording these measurements, astronomy and astrophysics have nothing more to say. If they draw any inferences from the measurements, these inferences cannot be considered as legitimate science. That is the positivist standpoint. The mental constructions that we make in collating and selecting and systematizing the measurement data, and the theories which we advance to explain why they should be so and not otherwise, are an unwarranted human intrusion on the scene. They are mere arbitrary inventions of human reason. They may be convenient, just as the habit of thinking in similes is a convenient help to the mind, but we have no right to put them forward as representing anything that really happens in nature.

All we know is the bare result of the sensory measurements and we have no right to attach an ulterior significance to these.

Supposing we say, with Ptolemy, that the earth is the fixed centre of the universe and that the sun and all the stars move around it; or supposing we say, with Copernicus, that the earth is a small particle of matter which is relatively insignificant in relation to the whole universe, turning on its axis once every twenty-four hours and revolving around the sun once in every twelve months—on the positivist principle the one theory is as good as the other, when considered from the scientific viewpoint. They are merely two different ways of making a mental construction out of sensory reactions to some outer phenomena; but they have no more right to be looked upon as scientifically significant than the mental construction which the mystic or poet may make out of his sensory impressions when face to face with nature. It is true that the Copernican theory of astronomy is more widely accepted; but that is because it is a simpler way of formulating a synthesis of sensory observations and it does not give rise to so many difficulties about astronomical laws as would arise from the acceptance of the Ptolemaic theory. Therefore Copernicus is not to be judged as a pioneer discoverer in the realms of science, no more than a poet is to be judged as a pioneer discoverer when he gives fanciful and attractive expression to sentiments that are known to every human breast. Copernicus *discovered* nothing. He only formulated, in the shape of a fanciful mental construction, a mass of facts that were already known. He did not add anything to the store of scientific knowledge already in existence. A tremendous mental revolution was caused by his theory and bitter battles were waged around it. For the logical consequence of it was to give an entirely different account of man's place in the universe from that generally held at the time by the religion and philosophy of Europe. But for the positivist scientist all the fuss and trouble made over the Copernican theory were quite as senseless, from the scientific point of view, as if one were to quarrel with the rapture of a contemplative who gazes on the Milky Way and ponders over the fact that each star in that Milky Way is a sun somewhat like ours and that each spiral nebula is again a Milky Way from which the light has taken many millions of years to reach our earth, while the earth itself, with its human race on it, sinks away into an insignificant speck which is hardly discernible in the boundless space.

Incidentally we must remind ourselves that to look at nature in this way is to look at it from the aesthetic and ethical standpoints. These, of course, have no direct relation to physical science. Therefore they are excluded. But in excluding them there is a fundamental difference between the attitude of the non-positivist and that of the positivist physicist. The ordinary scientist, who does not believe in the positivist attitude, admits the validity of the aesthetic standpoint and the ethical standpoint; but he recognizes these as belonging to another way of looking at nature. Such a way does not come within the province of physical science. On the other hand, the positivist does not admit any such values as real at all, even in other provinces than physical science. For him a beautiful sunset is merely a sequence of sensory impressions. Therefore, as I said at the beginning, as long as we logically pursue the positivist teaching we must exclude every influence of a sentimental, aesthetic or ethical character from our minds. We have to keep to the logical track. That is the indispensable guarantee of certainty which the positivist teaching has to offer. And here I may remind the reader once again that we are examining a system which has been put forward with the very laudable motive of furnishing a sure basis for the reliability of science. Therefore the whole position must be discussed entirely objectively and free from any polemical feeling.

In the positivist way of looking at nature sensory impressions are the primary data and therefore signify immediate reality. From this it follows that in principle it would be a mistake to speak of the senses themselves being deceived. What under certain circumstances can be deceptive are not the sensory impressions themselves but the conclusions we so often draw from them. If we plunge a straight stick into water and hold it slantwise, and notice the apparent bend at the point of immersion, we are not deceived by the sense of sight into thinking that the stick is thereby bent. There is an actual bending present as an optical perception; but that is quite a different thing from concluding that the stick itself is bent. The positivist will not allow us to conclude anything. We have a sensory impression of the part of the stick that is in water and a contiguous sensory impression of the part that is in air; but we have no right to say anything about the stick itself. The most that the positivist principle will allow us to say is that the stick looks "as if" it were bent. If we explain the whole phenomenon by saying that the light rays which are reflected in the air from the stick to the eye pass through a less dense medium than that through which the rays pass when reflected from the part of the stick immersed in water, and that therefore the latter are more strongly deflected, that way of stating the case is useful from many points of view but it is no closer to reality than to say that the senses perceive the stick "as if" it were bent.

The essential point here is that, from the standpoint of Positivism, both ways of stating the case are fundamentally of equal validity. And there would be no sense in attempting to judge their rival validities by asking how far one is more appropriate than the other, by appealing to the sense of touch to rectify the apparent anomaly of a stick which was straight in air being bent in water. In the positivist system there would be no meaning in a decision one way or another; because a strictly logical positivist science would have to be content with merely noting the sensory impressions and leaving the matter at that. We could say that the stick looks "as if" it were bent. In practice, of course, anything like a serious attempt at an all-round application of this "as if" theory would lead to ridiculous consequences. But here we are not testing the positivist theory by any such grounds. We are considering it on its own chosen ground of logical consistency, which is its bedrock foundation. It must stand or fall by the consequences that would result for physical science by the logical application of the positivist premises.

What I have said here in regard to the stick applies equally to all the surrounding objects of inanimate nature. In the positivist view a tree is nothing more than a complex of sense-impressions. We can see it grow. We can hear the rustle of its leaves and inhale the perfumes of its blossoms. But if we take away all these sensory impressions then nothing remains to correspond to what may be called the "tree in itself."

What holds good for the world of plant life must also have meaning for the animal world. We speak of this world as a special and independent realm of being, but that is solely because it is a convenient way of thinking and talking. If we tread on a worm it squirms. That we can see. But there would be no sense in asking if the worm suffers pain thereby. For a man can feel only his own pain and he cannot with any certainty of knowledge extend that same feeling to the animal world. To say that an animal suffers pain is an assumption based on a summary of various characteristics that correspond to what happens in our own case under similar circumstances. In the case of a worm we notice a squirming or shrugging. In the case of other animals we notice contortions of the face and body. These are analogous to what happens in our case under like conditions. And there are certain cries in the animal world which are analogous to the sounds we utter when we suffer pain.

When we come from the animal world to the world of human beings we find the positivist scientists making a clear distinction between one's own impressions and the impressions of others. One's own impressions are the sole reality and they are realities only for oneself. The impressions of another person are only indirectly knowable to us. As objects of knowledge they signify something fundamentally different from our own impressions. Therefore in speaking of them we are merely following the same sort of useful analogy as when we speak of the suffering of animals. But, in the strict positivist view, we have no reliable knowledge whatsoever of other people's impressions. Because they are not a direct sensory perception, they do not furnish a basis for the certainty of our knowledge.

It is quite clear that the positivist outlook cannot be accused of logical inconsistency. So long as we stick closely to its principles we do not find ourselves up against any contradiction. That is the strong point of the whole system. But when we come to apply it as the exclusive foundation on which scientific research can be carried on we shall find that the result would be of very significant import for physical science. If the scope of physical science extends no further than the mere description of sensory experiences, then strictly only one's own experiences can be taken as the object of such description; because only one's own experiences are primary data. Now it is clear that on the basis of a mere individual complex of experience not even the most gifted of men could construct anything like a comprehensive scientific system. So we are faced with the alternative of either renouncing the idea of a comprehensive science, which will hardly be agreed to even by the most extreme positivist, or to admit a compromise and allow the experiences of others to enter into the groundwork of scientific knowledge. But we should thereby, strictly speaking, give up our original standpoint, namely, that only primary data constituted a reliable basis of scientific truth. The sensory impressions of others are secondary and they are data for us only through the reports we have of them. This brings a new factor into play here, namely, the trustworthiness of oral and written information in scientific reports. Therewith we break at least one link of the logical chain which holds the positivist system together; for the foundational principle of the system is that only immediate perception can be considered as offering material for scientific certainty.

Let us, however, pass over this difficulty and let us assume that all reports furnished by scientific researchers are reliable or at least that we have an infallible means of excluding those which are unreliable. In this case it is obvious that the reports furnished by the numerous scientists who were and are acknowledged as honourable and reliable both in the past and to-day must be taken into scientific consideration; and there are no grounds whereon some should be excluded in favour of others. It would be quite wrong to devaluate the claims of any investigator on the grounds that his findings have not been corroborated by others.

If we should stick to this idea then it would be difficult to explain or to justify the conduct of physical science in regard to certain individual researchers. Let us take one instance as illustrative.

The so-called N-rays which were discovered by the French physicist, Blondlot, in the year 1903, and at that time studied on all sides, are to-day entirely ignored. Réné Blondlot, who was professor at the University of Nancy, was admittedly an excellent and reliable investigator. His discovery was for him an experience as great as that of any other physicist. We cannot say that he was fooled by his sense-perceptions; for in positivist physics, as we have seen, there is no such thing as delusion in sensuous perception. It would be only proper and right to look upon the N-rays as primary reality-data, something that directly struck the perception of one man. And if since the time of Blondlot and his school no man throughout all the years between has succeeded in reproducing them, that is no reason for saying—at least from the positivist standpoint—that they will not one day, under some special circumstances, yet again become discernible.

Under the positivist test we should have to agree that the number of those researchers whose findings are of value for physical science is indeed very small. We should have to admit only those who devote themselves specially to this science, because the discoveries which outsiders have made in this field are more or less insignificant. Moreover, we must from the outset exclude all theoretical physicists; for their experiences are restricted essentially to the use of pen, ink, and paper and abstract reasoning. And thus we have only the experimental physicists remaining, and in the first line only those who confine themselves to the operation of extremely sensitive instruments for special investigation. Therefore in the positivist hypothesis only a small roll of specially qualified physicists come into the picture when we speak of the contributions of those who have devoted themselves to the progress of physical science.

From this standpoint how are we to explain the extraordinary impression made and the revolution which was created in the world of international science by the findings, for instance, of Oersted, who detected the influence of a galvanic current on the compass needle, or of Faraday, who first discovered the effect of electromagnetic induction, or of Hertz, who discovered small electric sparks in the focus of his parabolic reflector by the use of the magnifying glass? How and why did these individual sensory impressions create such a furore and lead to such a world revolution in the theory and application of scientific methods? To this question the upholders of positivism can give only a roundabout and entirely unsatisfactory answer. They have to fall back upon the theory that these individual experiences, which were insignificant in themselves, merely opened up a viewpoint as a result of which other researchers were led to the discovery of a series of much greater and more portentous results. That is a rather lame answer but it illustrates very well the positivist position, because the upholder of positivism will admit nothing except a bald description of results experienced in research; and if we ask why it is that certain findings of a few obscure individuals, carried out under quite primitive conditions, had such an immediate and world-wide significance for all other physicists—that question has no meaning for physical science as viewed from the positivist standpoint.

The reason for taking up this striking attitude is quite easy to understand. Those who lean towards the discipline that I have been describing deny the idea and the necessity of an objective physical science which is independent of the actually experiencing and sense-perceiving investigator. They cling to this attitude because they are bound logically to acknowledge no other reality save that of the factual experience of the individual physicist. Now I think it is obvious here that if physical science as such were to accept this position, as the exclusive basis of its research, then it would find itself trying to support a huge structure on a very inadequate foundation. A science that starts off by predicating the denial of objectivity has already passed sentence on itself. Of what value to the world are the sensory impressions of a mere individual? Yet that is the foundation to which in the last analysis physical science is reduced in looking for a basis for its structure. This plot is entirely too small for such a building. It has to be extended by the addition of other ground. No science can rest its foundation on the dependability of single human individuals. And the moment we have made that statement we have taken a step which puts us off the logical pathway of the positivist system. We have followed the call of common sense. We have taken a jump into the meta-physical realm; because we have accepted the hypo-thesis that sensory perceptions do not of themselves create the physical world around us, but rather that

F

they bring news of another world which lies outside of ours and is entirely independent of us.

And thus we strike out the positivist *als-ob* (As-If) and attribute a higher kind of reality than that of mere description of immediate sensory impressions to the practical discoveries that have been already mentioned —Faraday's, etc. Once we take this step we lift the goal of physical science to a higher level. It is not restricted to the mere description of bare facts of experimental discovery; but it aims at furnishing an ever increasing knowledge of the real outer world around us.

At this point a new epistemological[1] difficulty enters. The basic principle of the positivist theory is that there is no other source of knowledge except within the restricted range of perception through the senses. Now there are two theorems that form together the cardinal hinge on which the whole structure of physical science turns. These theorems are: (1) *There is a real outer world which exists independently of our act of knowing,* and, (2) *The real outer world is not directly knowable.* To a certain degree these two statements are mutually contradictory. And this fact discloses the presence of an irrational or mystic element which adheres to physical science as to every other branch of human knowledge. The knowable realities of nature cannot be exhaustively discovered by any branch of science. This means that science is never in a position completely and exhaustively to explain the problems it has to face. We see in all modern

---

[1] Epistemology is the Science of the Nature of Knowledge.

scientific advances that the solution of one problem only unveils the mystery of another. Each hilltop that we reach discloses to us another hilltop beyond. We must accept this as a hard-and-fast irrefutable fact. And we cannot remove this fact by trying to fall back upon a basis which would restrict the scope of science from the very start merely to the description of sensory experiences. The aim of science is something more. It is an incessant struggle towards a goal which can never be reached. Because the goal is of its very nature unattainable. It is something that is essentially metaphysical and as such is always again and again beyond each achievement.

But if physical science is never to come to an exhaustive knowledge of its object, then does not this seem like reducing all science to a meaningless activity? Not at all. For it is just this striving forward that brings us to the fruits which are always falling into our hands and which are the unfailing sign that we are on the right road and that we are ever and ever drawing nearer to our journey's end. But that journey's end will never be reached, because it is always the still far thing that glimmers in the distance and is unattainable. It is not the possession of truth, but the success which attends the seeking after it, that enriches the seeker and brings happiness to him. This is an acknowledgment made long ago by thinkers of deepest insight, even before Lessing gave it the classic stamp of his famous phrase.

CHAPTER III

THE SCIENTIST'S PICTURE OF THE PHYSICAL UNIVERSE

THE ideal aim before the mind of the physicist is to understand the external world of reality. But the means which he uses to attain this end are what are known in physical science as *measurements*, and these give no direct information about external reality. They are only a register or representation of reactions to physical phenomena. As such they contain no explicit information and have to be interpreted. As Helmholtz said, measurements furnish the physicist with a sign which he must interpret, just as a language expert interprets the text of some prehistorical document that belongs to a culture utterly unknown. The first thing which the language expert assumes —and must assume if his work is to have any practical meaning—is that the document in question contains some reasonable message which has been stated according to some system of grammatical rules or symbols. In the same way the physicist must assume that the physical universe is governed by some system of laws which can be understood, even though he cannot hold out to himself the prospect of being able to understand them in a comprehensive way or to discover their character and manner of operation with anything like a full degree of certitude.

Taking it, then, that the external world of reality

is governed by a system of laws, the physicist now constructs a synthesis of concepts and theorems; and this synthesis is called the scientific picture of the physical universe. It is a representation of the real world itself in so far as it corresponds as closely as possible to the information which the research measurements have supplied. Once he has accomplished this the researcher can assert, without having to fear the contradiction of facts, that he has discovered one side of the outer world of reality, though of course he can never logically demonstrate the truth of the assertion.

If we consider the efforts that have been made by physicists, ever since the days of Aristotle, to describe the external universe, I think we need have no hesitation in expressing unqualified admiration for the extraordinary degree of perfection achieved in this respect by the inventive mind of the scientific researcher. From the positivist standpoint, of course, this idea of constructing a scientific picture of the physical universe—this continual striving after a knowledge of external reality—is something foreign and meaningless. For where there is no outer object there is nothing that can be portrayed or described.

The chief quality to be looked for in the physicist's world-picture must be the closest possible accord between the real world and the world of sensory experience. What is taken in through the senses is the first material that the physicist has to work upon. And the first process which this raw material must undergo is one of elimination and refinement. From the whole complex of sensory data everything must be cut away and discarded which may have arisen from the subjective constructive tendencies of the sensory organs themselves. And, furthermore, everything must be eliminated which can be attributed to the accident of special circumstances. In this latter connection attention must be paid to the fact that measuring instruments may affect the results that are being arrived at during the process of observation. That is all the more likely to be the case in the observation of minutiae.

Supposing all the above conditions to have been verified, then the physicist's picture of the external universe has only one further requirement to fulfil. Throughout its whole composition it must be free from everything in the nature of a logical incoherence. Otherwise the researcher has an entirely free hand. He may give rein to his own spirit of initiative and allow the constructive powers of the imagination to come into play without let or hindrance. This naturally means that he has a significant measure of freedom in making his mental constructions; but it must be remembered that this freedom is only for the sake of a specific purpose and is a constructive application of the imaginative powers. It is not a mere arbitrary flight into the realms of fancy.

The physicist is bound, by the very nature of the task in hand, to use his imaginative faculties at the very first step he takes. For the first stage of his work must be to take the results furnished by a series of experimental measurements and try to organize these under one law. That is to say, he must select according to a plan which will in the first instance be hypothetical and therefore a construction of the imagination. And when he finds that the given results will not fit into one plan he discards it and tries another. This means that his imaginative powers must always be speculating on the significance of the data which have been furnished through experimental measurements. He is in the same position as a mathematician who is presented with a number of single points that have to be joined together with a curve. The closer together and the more numerous these points are, the more innumerable will be the possible kinds of alternative curves that present themselves to the mind. We meet with practically the same task when we follow the movements of a sensitive registering instrument which is designed to mark only one independent and definite curve, such as the temperature curve; for we find that this curve is never sharply defined but is always a more or less broad stroke in which an endless number of sharp curves find their place.

As to how one may reach a decision in the midst of this uncertainty, no general rule of procedure can be laid down. One must simply choose a definite line of thought. And that line of thought ought to be directed towards founding, on the basis of a selected combination of ideas, a hypothesis in the light of which we can outline the curve we are seeking for, and outline it in such a way that it will have a clarity and definiteness of its own that distinguishes it from the numberless other curves intruding on the scene. In other words, where the spectrum shows, for instance, a diversified picture, and where we are seeking for the cause of only one element in that picture, we have to imagine a number of hypothetical causes and examine them one after the other until we hit upon something that will accord with a certain series of results that are pictured on the spectrum. The line of thought which leads to these various alternatives has its origin entirely outside the ambit of logic. In order to formulate this kind of hypothesis the physicist must possess two characteristics. He must have a practical knowledge of his whole field of work and he must have a constructive imagination. This means that, in the first place, he must be acquainted with other kinds of measurements besides the one that he is actually using. And, in the second place, he must have the knack of combining under one viewpoint the results obtained through two different kinds of measurement.

Every hypothesis that is productive of results has its origin in some fortunate juxtaposition of two different ways in which observations have presented themselves. We see this truth very clearly illustrated in the famous historical cases that have led to epoch-making discoveries.

When Archimedes had noted the loss of weight registered by his own body in water he connected this fact with the loss of weight which various other bodies would undergo on being placed in water, and thus he arrived at a means of finding out the specific

gravities of various metals. This came into his head
one day while in his bath and meditating how he
could assay the golden crown of the King of Syracuse,
which was suspected of containing a silver alloy,
though it purported to be of pure gold. Applying the
experience of his own loss of weight in the bath, it
struck him that the excessive bulk occasioned by the
alloy could be detected by putting the crown, and
equal weights of gold and silver, separately in a vessel
of water and measuring the difference of the overflow.
Newton noticed the movement of an apple falling
from a tree in his orchard and he connected that
observation with the motion of the moon in relation
to the earth. Einstein observed the state of a gravita-
ting body in a fixed box, and considered this in juxtaposition with the state of a body free from gravitation
in a box subjected to a process of upward acceleration.
Niels Bohr associated the orbital rotation of the
electrons around the nucleus of an atom with the
movement of planets around the sun. All these
combinations were productive of famous results.
Indeed it would be an interesting mental exercise if
one were to take as many as possible of the hypotheses
which have proved significant of results in the pursuit
of physical science and then try to discover the respec-
tive combinations of ideas to which the hypotheses
owed their origin. But the task would be a difficult
one because, generally speaking, creative master
minds have felt a personal aversion from the idea of
unfolding before the public gaze those delicate threads
of thought out of which their productive hypotheses

were woven, and the myriad other threads which failed to be interwoven into any final pattern.

The utility of an hypothesis, once it has been put forward, can be tested only by following out the logical results that flow from its application. This has to be done in a purely logical—and primarily mathematical —way, whereby the hypothesis is used as a starting-point and as complete a theoretical system as possible is developed from it. Once the theoretical system has thus been fully developed it will be put to the test of the measurements which have been furnished by factual experiment. According as the system closely corresponds with these measurements we can judge whether the hypothesis from which we started was or was not successfully chosen.

Such being the actual method of procedure adopted by the physicist, we can understand at once how it is that the progress of physical science does not follow a regular curve of development, which might mark an all-round process of increasing depth and precision in the knowledge we are gaining of the external world. It is rather a zig-zag pattern that the curve of scientific progress follows; indeed I might say that the forward movement is of an explosive type, where the rebound is an attendant characteristic of the advance. Every applied hypothesis which succeeds in throwing the searchlight of a new vision across the field of physical science represents a plunge into the darkness; because we cannot at first reduce the vision to a logical state-ment. Then follows the birth-struggle of a new theory. Once this has seen the light of day it has to go forward willy-nilly until the stamp of its destiny is put on it when the test of the research measurements is applied. If the hypothesis survives this test, then it advances in prestige and acceptance and the theory arising from its application develops and expands to a more and more comprehensive ambit.

But, on the other hand, if the application of research measurements places difficulties to the viability of our hypothesis, then fears and misgivings and critical birth conditions set in. But these are the signs of the breaking up of old acceptances and the bringing to birth of a new hypothesis. The task of the latter will be to push forward to the solution of the crisis out of which it was born, and to construct a new theory which will preserve what was genuine in the old order of things, while correcting and discarding the mistakes. So, in the everlasting interplay of change succeeding upon change, the knowledge which physical science brings to us comes at one time with hesitating step and at another with a forward bound, in its way towards the discovery of the real external universe.

This has been a regularly recurrent feature throughout the historical development of physical science. Take the case of the Lorentzian theory of electrodynamic motion. The conflicts and contradictions which were set afoot by the application of actual research measurements in this case are well known; but only those who have closely followed the thorny path of the Lorentzian theory step by step can rightly appraise the relief which came to hand when the

Relativity hypothesis was first established. An almost exactly similar experience has been encountered in the history of the quantum theory; but in this latter case the crisis is not yet entirely passed.

It has already been said that in the statement of any hypothesis the author of it has a free hand from the very start. He has full and free choice of the concepts and theorems which he will employ in framing his synthesis, provided of course that there is no logical contradiction between them. It is not true, as has often been stated in physicist circles, that in the exposition of an hypothesis the explorer must draw the material for his ideas solely and strictly from those original data which have been definitely furnished by the results of the research measurements. This would mean that the formative concepts which give shape to an hypothesis must be strictly independent of all theoretical origin. That is not so. For, on the one hand, every hypothesis—as a factor in the picture of the external universe presented by the physicist —is a product of the freely speculating human mind; and, on the other hand, there are no physical formulae whatsoever which are the immediate results of research measurements. The opposite is the case. Every measurement first acquires its meaning for physical science through the significance which a theory gives it. Anybody who is familiar with a precision laboratory will agree that even the finest and most direct measurements—such as those of weight and current—have to be corrected again and again before they can be employed for any practical purpose. It is obvious that these corrections cannot be suggested by the measurement process itself. They must first be discovered through the light which some theory or other throws upon the situation; that is to say, they must arise from an hypothesis.

The truth of the whole matter is that the inventor of an hypothesis has unlimited scope in the choice of whatever means he may deem helpful to his ultimate purpose. He is not hindered by the physiological tendencies towards constructive picturing which are a feature of the activity of his own sense-organs. Nor is he restricted by the guiding hand of his physical measuring-gear. With the eye of the spirit he penetrates and supervises the most delicate processes that unfold themselves in the pattern of the physical universe which unrolls before him. He follows the movements of every electron and watches the frequency and form of every wave. He even invents his own geometry as he goes along. And so with his spiritual working-gear, with these instruments of ideal exactitude, he takes a personal part, as it were, in every physical process that happens before him. And all this is for the purpose of pushing through these difficult thought experiments—which are a factor of every research process —to the final establishment of conclusions that will be of wide application. Naturally all such conclusions have, at the outset of their statement, nothing to do with the real research measurements. And therefore an hypothesis can never be declared true or false in the light of such measurements. All that can be asked about it is how far it reaches or falls short of serving some practical purpose or other.

And now we come to the other side of the picture. This ideal clear-sightedness of the spiritual eye, in seeing behind the various processes of physical nature, is due exclusively to the fact that the nature of the physical world in this case is something that is fashioned by the mind of the observer himself. As long as this world of his intuitive construction remains an hypothetical world, the creator has full knowledge of it, and full dominion over it and can shape it what way he will; because as far as concerns reality it has as yet no value. The first value comes the moment the theoretical system on which this hypothetical world has been planned is brought into touch with actual results that have been furnished through research measurements.

Now, a merely physical measuring process tells us just as little about the account which we are to give of the physical universe as it does about the reality of that universe itself. Indeed the process of research measurement rather represents a happening in the sense-organs of the researcher in relation to the happening that takes place in the apparatus that he is using. All that can be definitely said about this relation in respect of outer reality is that there is some connection or other between them. The measurement itself gives no immediate results that have a meaning of their own. And it is the task of science to try to establish the meaning of the above-mentioned connection, quite as much as it is the task of the scientific explorer to carry out the actual physical measurements themselves. The former task can be accomplished only by the speculative mind of the researcher.

The epistemological difficulties which have arisen in the sphere of theoretical physics through the development of the quantum theory seem to be due to the fact that the bodily eye of the measuring physicist has been identified with the spiritual eye of the speculative scientist. As a matter of fact, the bodily eye, being part of the physical process of nature itself, is the object rather than the subject of scientific exploration. For as every act of research measurement has a more or less causal influence on the very process that is under observation, it is practically impossible to separate the law that we are seeking to discover behind the happening itself from the methods that are being used to bring about the discovery.

It is true that where there is question of natural phenomena in the lump, such as a group of atoms taken together, the method of measurement is not so likely to influence the course of the events observed. And it is for this reason that in the earlier stages of physical science, which are now called the classical physics, the opinion held sway that the actual measurement itself furnishes a direct glimpse into the real happenings of nature. But in this assumption, as we have already seen, there was a fundamental mistake which is the counterpart of the positivist error, namely, the paying of attention solely to the results given by experimental measurements and entirely ignoring the inner reality of natural processes. Yet while we recognize this as a mistake on the one side we must also realize on the other side that if we are to abandon the measuring method we have no way of coming into touch with the real happenings themselves. But when we are faced with the indivisible quantum of action, the limit is laid with mathematical accuracy, beyond which the most delicate physical measurement is unable to give a satisfactory answer to questions connected with the individual behaviour of the more minute processes. The result is that the problem of these infinitesimal processes has no longer a meaning for purely physical research. Here we come to the point where such problems have to be dealt with by the speculative reason. And it is in this abstract way that they must be taken into account in our attempt to complete the physicist's picture of the universe and thus bring us nearer to the discovery of external reality itself.

Taking a glance backwards over the road along which physical science has hitherto advanced we must admit that further progress will depend essentially on the development and wider application of our methods of measurement. Thus far I am at one with the positivist outlook. But the difference between us is that positivism holds research measurement, through sensory perception, as the be-all and end-all of the processes through which physical science advances, whereas I hold that the study of physical realities treats measurement results as a more or less intricate complex representing the registration of reactions to happenings in the external world, the accuracy of which registration is relatively dependent on what takes place in the registering instruments themselves and in the interpretative sensory organs of the researcher. The adequate analysis and correction of this complex report is one of the chief functions of scientific research. Therefore from the results that are given by experimental measurements we must choose those which will have a practical bearing on the object of our inquiry, because each particular attempt at discovering reality in the physical universe represents a special form of a certain question which we put to nature.

Now you cannot put a reasonable question unless you have a reasonable theory in the light of which it is asked. In other words, one must have some sort of theoretical hypothesis in one's mind and one must put it to the test of research measurements. This is why it often happens that a certain line of research has a meaning in the light of one theory but not in that of another. And very often the significance of a question changes when the theory in the light of which it is asked has already changed.

Let us take for example the transmutation of some common metal, such as quicksilver, into gold. For those who lived in the days of alchemists this problem had a very important significance and innumerable researchers sacrificed their means and their life's efforts in an attempt to solve it. The problem lost all its meaning, and came to be looked upon as a fool's pursuit, when the dogma of the intransmutability of

G

atoms was introduced. But now once again, since Bohr has put forward his theory that the gold atom is different from the quicksilver atom only by the lack of one single electron, the problem has become so vital that it is being newly worked upon with the use of the most modern research methods. Here again one sees the truth of the old adage, that experience is the pathfinder of scientific study. When intelligently worked out, even the most useless experiments may result in opening up a way to the most important discoveries.

It was thus that those more or less planless attempts to make gold opened up the way to the introduction of scientific chemistry. So too from the unsolved problem of the perpetuum mobile finally arose the principle of the conservation of energy. And the long series of vain attempts to measure the movement of the earth led at last to the suggestion of the conditions from which the theory of relativity arose. Experimental and theoretical adventures in science are always interdependent. The one cannot progress without the other.

It often happens that when a new advance in theoretical science has definitely established itself certain problems connected with it are branded as meaningless. Not only that, but attempts are also sometimes made to prove such problems meaningless on *a priori* grounds. That is a delusion. In itself neither the absolute motion of the earth—that is to say, the motion of the earth in relation to the light-ether—nor the absolute Newtonian space, is meaningless, as has often been declared by popular exponents of the relativity theory. The former problem is meaningless only when you introduce the special theory of relativity, and the latter is meaningless only when you introduce the general relativity theory.

So when we look back over the centuries we see that doctrines of the interpretation of nature, which were held as sound and good for their time, fell from honour when faced with the light of some new scientific theory. They served their day and then they passed. And though succeeded by more scientifically enlightened dogmas we must remember that those old theories had sense and meaning for their age, as other theories will have had sense and meaning for our time; until another day comes when newer theories will arise to take their place.

The Law of Causality was unanimously accepted until recent times as a fundamental principle in scientific research. But now a battle of opinions is being waged around it. Does the principle of causality, as hitherto believed, hold good in all its force for every physical happening? Or has it only a summary and statistical significance when applied to the finer atoms? This question cannot be decided by referring it to any epistemological theory or by putting it to the test of research measurements. In his attempt to build up his hypothetical picture of the external universe the physicist may or may not, just as he likes, base his synthesis on the principle of a strict dynamic causality or he may adopt only a statistical causality. The important question is how far he gets with the one or the other. And that can be answered only by choosing provisionally one of the two stand-points and studying the conclusions which can be logically derived from the adoption of that standpoint, just as we did when dealing with Positivism.

In principle it does not matter which of the two standpoints is chosen first. In practice one will naturally choose that which promises to turn out more satisfactory in its logical results. And here I must definitely declare my own belief that the assumption of a strict dynamic causality is to be preferred, simply because the idea of a dynamically law-governed universe is of wider and deeper application than the merely statistical idea, which starts off by restricting the range of discovery; because in statistical physics there are only such laws as refer to groups of events. The single events, as such, are introduced and recognized expressly; but the question of their law-governed sequence is declared senseless on *a priori* grounds. That way of procedure appears to me to be highly unsatisfactory. And I have not been able to find the slightest reason, up to now, which would force us to give up the assumption of a strictly law-governed universe, whether it is a matter of trying to discover the nature of the physical, or the spiritual, forces around us.

It is obvious, of course, that no strictly causal connection can be deduced from a succession of experimental experiences. Between these experiences as they succeed one another we can establish only a statistical relation. Even the acutest measurements are subject to accidental and uncontrollable mistakes.

An experimental observation presents, as we have seen, a complex result made up of several different elements. And even though each element were the direct causal consequence of one other single element, yet we cannot treat this original element as strictly causal in the experiment, independently of the others; because diversified results may follow from the combination in which each elemental factor is applied.

And here a question arises which seems to set a definite impassable limit to the principle of strict causality, at least in the spiritual sphere. This question is of such urgent human interest that I think it will be well if I treat it here before I come to a close. It is the question of the freedom of the human will. Our own consciousness tells us that our wills are free. And the information which that consciousness directly gives us is the last and highest exercise of our powers of understanding.

Let us ask for a moment whether the human will is free or whether it is determined in a strictly causal way. These two alternatives seem definitely to exclude one another. And as the former has obviously to be answered in the affirmative, so the assumption of a law of strict causality operating in the universe seems to be reduced to an absurdity in at least this one instance. In other words, if we assume the law of strict dynamic causality as existing throughout the universe, how can we logically exclude the human will from its operation?

Many are the attempts that have been made to solve this dilemma. The purpose which in most cases they have set themselves has been to establish an exact limit beyond which the law of causality does not apply. Recent developments in physical science have come into play here, and the freedom of the human will has been put forward as offering logical grounds for the acceptance of only a statistical causality operative in the physical universe. As I have already stated on other occasions, I do not at all agree with this attitude. If we should accept it, then the logical result would be to reduce the human will to an organ which would be subject to the sway of mere blind chance. In my opinion the question of the human will has nothing whatsoever to do with the opposition between causal and statistical physics. Its importance is of a much more profound character and is entirely independent of any physical or biological hypothesis.

I am inclined to believe, with many famous philosophers, that the solution of the problem lies in quite another sphere. On close examination, the above-stated alternative—Is the human will free or is it determined by a law of strict causality?—is based on an inadmissible logical disjunction. The two cases opposed here are not exclusive of one another. What then does it mean if we say that the human will is causally determined? It can only have one meaning, which is that every single act of the will, with all its motives, can be foreseen and predicted, naturally only by somebody who knows the human being in question, with all his spiritual and physical characteristics, and who sees directly and clearly through his conscious and subconscious life. But this would mean that such a person

would be endowed with absolutely clear-seeing
spiritual powers of vision; in other words he would be
endowed with divine vision.

Now, in the sight of God all men are equal. Even
the most highly gifted geniuses, such as a Goethe or
a Mozart, are but as primitive beings the thread of
whose innermost thought and most finely spun
feelings is like a chain of pearls unrolling in regular
succession before His eye. This does not belittle the
greatness of great men. But it would be a piece of
stupid sacrilege on our part if we were to arrogate to
ourselves the power of being able, on the basis of our
own studies, to see as clearly as the eye of God sees
and to understand as clearly as the Divine Spirit
understands.

The profound depths of thought cannot be pene-
trated by the ordinary intellect. And when we say that
spiritual happenings are determined, the statement
eludes the possibility of proof. It is of a metaphysical
character, just as the statement that there exists an
outer world of reality. But the statement that spiritual
happenings are determined is logically unassailable,
and it plays a very important rôle in our pursuit of
knowledge, because it forms the basis of every attempt
to understand the connections between spiritual
events. No biographer will attempt to solve the question of the motives that govern the acts of his hero
by attributing these to mere chance. He will rather
attribute his inability to the lack of source materials
or he will admit that his own powers of spiritual
penetration are not capable of reaching down into

the depths of these motives. And in practical everyday life our attitude to our fellow beings is based on the assumption that their words and actions are determined by distinct causes, which lie in the individual nature itself or in the environment, even though we admit that the source of these causes cannot be discovered by ourselves.

What do we then mean when we say that the human will is free? That we are always given the chance of choosing between two alternatives when it comes to a question of taking a decision. And this statement is not in contradiction with what I have already said. It would be in contradiction only if a man could perfectly see through himself as the eye of God sees through him; for then, on the basis of the law of causality, he would foresee every action of his own will and thus his will would no longer be free. But that case is logically excluded; for the most penetrative eye cannot see itself, no more than a working instrument can work upon itself. The object and subject of an act of knowing can never be identical; for we can speak of the act of knowing only when the object to be known is not influenced by the action of the subject who initiates and performs the act of knowing. Therefore the question as to whether the law of causality applies in this case or in that is in itself senseless if you apply it to the action of your own will, just as if somebody were to ask whether he could lift himself above himself or race beyond his shadow.

In principle every man can apply the law of causality

to the happenings of the world around him, in the
spiritual as well as in the physical order, according to
the measure of his own intellectual powers; but he
can do this only when he is sure that the act of applying
the law of causality does not influence the happening
itself. And therefore he cannot apply the law of
causality to his own future thoughts or to the acts of
his own will. These are the only objects which for the
individual himself do not come within the force of
the law of causality in such a way that he can understand its play upon them. And these objects are his
dearest and most intimate treasures. On the wise
management of them depend the peace and happiness
of his life. The law of causality cannot lay down any
line of action for him and it cannot relieve him from
the rule of moral responsibility for his own doings;
for the sanction of moral responsibility comes to him
from another law, which has nothing to do with the
law of causality. His own conscience is the tribunal
of that law of moral responsibility and there he will
always hear its promptings and its sanctions when he
is willing to listen.

It is a dangerous act of self-delusion if one attempts
to get rid of an unpleasant moral obligation by claiming
that human action is the inevitable result of an inexorable law of nature. The human being who looks upon
his own future as already determined by fate, or the
nation that believes in a prophecy which states that
its decline is inexorably decreed by a law of nature,
only acknowledges a lack of will power to struggle and
win through.

And so we arrive at a point where science acknowledges the boundary beyond which it may not pass, while it points to those farther regions which lie outside the sphere of its activities. The fact that science thus declares its own limits gives us all the more confidence in its message when it speaks of those results that belong properly to its own field. But on the other hand it must not be forgotten that the different spheres of activity of the human spirit can never be wholly isolated from one another; because there is a profound and intimate connection between them all.

We started on the territory of a special science and have dealt with a series of problems that are of a purely physical character; but these have led us from the world of mere sense-perception to the real metaphysical world. And this world faces us with the impossibility of knowing it directly. It is a land of mystery. It is a world whose nature cannot be comprehended by our human powers of mental conception; but we can perceive its harmony and beauty as we struggle towards an understanding of it. And here on the threshold of this metaphysical world we have been brought face to face with the highest question of all, that of the freedom of the human will. It is a question which each one must meditate upon for himself if he thinks at all seriously on what the meaning of this life may be.

CHAPTER IV

CAUSATION AND FREE WILL

STATEMENT OF THE PROBLEM

THIS is one of man's oldest riddles. How can the independence of human volition be harmonized with the fact that we are integral parts of a universe which is subject to the rigid order of nature's laws?

At first sight these two aspects of human existence seem to be logically irreconcilable. On the one hand we have the fact that natural phenomena invariably occur according to the rigid sequence of cause and effect. This is an indispensable postulate of all scientific research, not merely in the case of those sciences that deal with the physical aspects of nature, but also in the case of the mental sciences, such as psychology. Moreover, the assumption of an unfailing causal sequence in all happenings is the basis on which our conduct of everyday life is regulated. But, on the other hand, we have our most direct and intimate source of knowledge, which is the human consciousness, telling us that in the last resort our thought and volition are not subject to this causal order. The inner voice of consciousness assures us that at any given moment we are capable of willing this or that alternative. And the corollary of this is that the human being is generally held responsible for his own actions. It is on this assumption that the ethical dignity of man is based.

How can we reconcile that dignity with the principle

of causation? Each one of us is an integral part of the world in which we live. If every other event in the universe be a link in the causal chain, which we call the order of nature, how can the act of human volition be looked upon as independent of that order? The principle of causation is either universally applicable or it is not. If not, where do we draw the line, and why should one part of creation be subject to a law that of its nature seems universal, and another part be exempted from that law?

Among all civilized races the profoundest thinkers have tackled this problem and have suggested innumerable solutions. I have no intention of adding to the sum-total here. My reason for taking up the question in connection with my own science is that the controversy has now entered the scientific field. From suggestions which have been made as to the inapplicability of the causal principle to certain types of research in physical science extensive conclusions have been drawn and the age-old controversy is now being waged more bitterly than ever.

After all the thought that has been expended on it, since man first began to reason over his place in the universe, one might justifiably assume that the problem of causation would be nearer to a solution now than formerly, even if we grant that a complete and final solution is impossible, from the very nature of the question itself. And we might reasonably expect that at this stage of the controversy the disputants would at least be in agreement as to the nature of the fundamental issues under discussion. But the opposite is the case. Nowadays it is not merely the problem itself that is debated; but even the very basic ideas involved in it are called into question—ideas such as the meaning of the concept of causality in itself and epistemological questions regarding the objects which should be considered to be within the legitimate scope of human knowledge, the difference between objects that are sensuously perceptible and objects that are outside this range and other such questions. All this quarrelling over fundamentals has added to the confusion.

The protagonists are mainly divided into two schools. One school is interested in the question chiefly from the viewpoint of the advancement of knowledge, holding that the principle of strict causation is an indispensable postulate in scientific research, even including the sphere of mental activity. As a logical consequence of this attitude, they declare that we cannot except human activity in any shape or form from the universal law of causation. The other school is more concerned with the behaviour of human beings and with the sense of human dignity, which feels that it would be an unwarrantable degradation if human beings, including even the mentally and ethically highest specimens of the race, were to be considered as inanimate automata in the hands of an iron law of causation. For this school of thinkers the freedom of the will is the highest attribute of man. Therefore we must hold, they say, that the law of causation is excluded from the higher life of the soul, or at least that it does not apply to the conscious mental acts of the higher specimens of humanity.

Between these two schools there is a great number of thinkers who will not go the whole distance in either direction. They feel in a certain sense that both parties are right. They will not deny the logical validity of the one position nor the ethical validity of the other. They recognize that in the mental sciences the principle of causation, as a basis of scientific research, is nowadays being pushed far beyond the borders of inanimate nature and with advantageous results. Therefore they will not deny the play of causality in the mental sphere, though they would like to erect a barrier somewhere within that sphere and entrench the freedom of human volition behind that barrier.

Among those who do not belong to either of the extreme schools perhaps I ought also to mention those scientists who are against the universal application of the principle of causality in physical science. They hold that it is inapplicable to the natural phenomena that are studied in quantum physics. But most of the scientists who hold this do not question the universal validity of the principle in itself. Still the attitude must be mentioned here; because, though it does not form anything like a school of thought, it indicates a tendency. And inasmuch as that tendency has been exploited by popularizers, who speak of spontaneity in the inner workings of nature, it deserves to be dealt with, if for no other purpose than to keep the lines of communication clear between serious science and the seriously thinking public.

As to the general controversy itself, if it did not

affect our approach to physical science physicists as such would not have to concern themselves with the matter. But the controversy now affects the very basic method on which scientific research is carried on. If the basis of causation be not valid, then how can the decisions arrived at on this basis be considered as reliable? Therefore the controversy affects the general claim to reliability which natural science puts forward. That is the reason why I am discussing it here as a physicist, in the hope that what I have to say may help to keep clear the grounds on which my own branch of science rests its claim to reliability.

Let us first consider the problem under its general aspect. What is the significance of the concept underlying the expression *Law of Causation*? In everyday life we are familiar with the idea of a *cause* and, like so many everyday things, we imagine that this idea is the simplest thing in the world to explain. Common sense and daily experience show us that all things and events are the products of other things and events. We say of what happens before our eyes that it is the effect of something else and we call that something else the cause, realizing at the same time that several causes may have contributed to bring about one and the same effect. On the other hand, we realize that effects themselves may be the cause of subsequent events.

When we find ourselves face to face with an event which we cannot possibly refer to any cause or series of causes, and which lies outside the range of all the causes that we are familiar with, then what happens?

Is it perfectly certain and necessary for human thought that for every event in every instance there must be a corresponding cause? Would the thought involve a logical contradiction that in this or that case the event has absolutely happened of itself and has no causal relation whatsoever to any other event? Of course the answer is in the negative; for it is very easy to *think* of an event as having no explanatory cause whatsoever. In such cases we speak of miracles and wonders and magic. And the simple fact that there exists a whole range of literature whose scenes are laid in wonderland is proof in itself that the concept of strict causality is not an inherent necessity of human thought. Indeed the human mind finds little difficulty in thinking of everything in the world as turning topsy-turvy. We can say to ourselves that to-morrow the sun may rise in the west, for a change. We can say to ourselves that a miracle of nature may occur, contrary to all the known laws of nature. We can think of the Niagara Falls for instance as shooting upwards, though this would be impossible in the world of reality. I can think of the door of my room in which I am now writing as opening of its own accord. And I can think of historical personages as entering the room and standing beside my table. In the world of reality to talk of such events may be meaningless and we may call them impossible, at least in our everyday way of reasoning. But we must distinguish this kind of impossibility from a logical impossibility, such as the idea of a square circle or that the part of something is greater than the whole, for no matter what efforts we make to think such things we cannot think them, inasmuch as they entail an inner contradiction. We can think of a part and we can think of the whole to which it belongs but we cannot think of the part as greater than the whole. This kind of impossibility is inherent in the nature of human thought itself, whereas the idea of something happening outside the range of causation is quite logically coherent.

Thus from the outset we can be quite clear about one very important fact, namely, that the validity of the law of causation for the world of reality is a question that cannot be decided on grounds of abstract reasoning. But reality, no matter what may be said to the contrary, is only a particular and small section of that immense sphere over which human thought can range. This is true even though our powers of imagination have always to take their cue from some real experience. Indeed experience is for us the starting-point of all thought; but we possess the gift of going beyond reality in thought. And were it not for this faculty of the imaginative intellect we should have no poetry and no music and no art. Indeed it is one of the highest and most precious gifts that man possesses, this power of lifting himself in thought into the realms of light whenever the weight of everyday life presses upon him and makes itself intolerable.

The creations of art are similar to those of science at least to the extent that scientific research, in the strictest sense of the term, could never advance without the creative force of the imaginative intellect. The man who cannot occasionally imagine events and conditions of existence that are contrary to the causal principle as he knows it will never enrich his science by the addition of a new idea. And this power of thinking beyond the range of causation is a prerequisite not only for the construction of hypotheses but also for the satisfactory co-ordination of results that have been arrived at through scientific research. It is the imaginative vision that puts forward a hypothesis. Then comes experimental research to put the hypothesis to its test. The results immediately arrived at through experiment have to be co-ordinated so as to form the basis of a theory, in the hope of discovering the laws of nature underlying the phenomena that have been studied. This work again calls the imaginative powers into play and further experiment puts the laws thus constructed to their final critical test.

To show how the scientific mind must necessarily imagine alternative happenings that lie outside the actual range of causation, when it is seeking to establish its conclusions, let us take a simple example from natural science as an illustration. Let us think of a ray of light coming to us from some distant star. Or indeed we can think of it as coming from some nearer source, such as an electric lamp. But let us think of it as passing through many transparent media of different nature and different densities, such as air, glass, water, etc., before it finally reaches the eye. What route will the light choose in coming from its point of origin to the eye of the observer? Generally speaking, this will not be a straight line; because when light passes through one medium after another its direction is bent from the direction of the line of entry. We are all familiar with this phenomenon in the case of a stick put into water. The line of light coming from the stick to the eye is bent at the point of emergence. And so the line of transmission for a beam of light coming from a distant source to the eye will be bent in each of the different transparent media through which it passes; so that its course will be zig-zag, according to the number and varying densities of the media. Even in the atmosphere itself the line which a ray of light follows is quite irregular, because the atmosphere possesses different powers of deflection at different heights.

Now, can we get any formula which states the actual route which our imaginary ray of light follows? We can. The answer is very definite. It is contained in that remarkable law of nature according to which a ray of light leaving a distant source will always choose, from the many alternative routes at its disposal, that route which will bring it to the eye of the observer within the shortest time, allowing for the fact that the light has to pass through the different media at different rates of speed. This is called the Principle of Quickest Arrival. And it has been a very useful principle in scientific research. But it would have no meaning whatsoever were we not in a position to imagine other alternative ways through which the light might travel, though in reality it does not travel along these ways and therefore they are causally impossible, in the sense that light cannot actually come by any other route. All the alternative routes that we may imagine are possible only in the abstract realm of the brain. They are impossible in the reality of nature. It is as if the light possessed a certain amount of intelligence and acted by the necessity of its own nature on the laudable principle of accomplishing its task in the quickest possible time. Therefore it has not the opportunity to dally and try out alternative ways, for it has to decide at once on the quickest way.

We have other similar cases in natural science such, for instance, as virtual motions which do not obey dynamical laws and therefore in the causal sense are impossible. But all these fanciful constructions play a very important rôle in theoretical science. They are employed as very useful instruments of thought in the carrying out of researches and the construction of theories. Therefore they certainly do not involve any contradiction of the laws of thought itself.

Once we have decided that the law of causality is by no means a necessary element in the process of human thought, we have made a mental clearance for the approach to the question of its validity in the world of reality. Now in the first place let us ask what is meant by the term, Causation? We might mean by it a regular interrelation between effects that follow one another in time. But we can at once ask whether this relation be founded in the nature of things themselves, or is it totally, or partly, a product of the imaginative faculty? Might it not be that mankind originally developed this concept of causation to meet the necessities of a practical life, but afterwards found that if men were to confine themselves to an outlook exclusively based on this principle life would then turn out to be unbearable? We need not delay here to discuss the various philosophical aspects of these questions. For our present purpose it is much more important to ask whether the causal connection between events must be considered as absolutely complete and always unbroken or are there events in the world which do not enter the chain as connecting-links?

Let us first see whether this question can be settled by a systematic application of deductive reasoning. As a matter of fact some of the most famous philosophers in the history of human thought have produced solutions of the causal problem which were based on purely abstract grounds. They took their first stand on the axiom *ex nihilo nihil fit*, that nothing comes from nothing, in other words that no event in the world holds in itself an adequate explanation of its own existence. Reasoning back from this standpoint the philosophers of what is generally called the rationalist school established as a logical necessity the existence of a Supreme Cause. This Supreme Cause is the God of Aristotle and the scholastic philosophers. As a logical consequence of the line of reasoning thus adopted it was necessary to attribute to this Godhead the possession in their plenitude of all the perfections that are present in the world. If there be an actually existent Supreme Cause outside of the world, who is the Creator of the world and the Creator of all things in the world, then man can deduce the nature of this Supreme Cause only through a study of His handiwork. From this one can easily see that the nature to be attributed to that Supreme Cause must necessarily depend upon man's outlook on created things. In other words the concept of the Divinity in this case must take its colour from the world outlook either of the individual philosopher in question or of the particular cultural background to which he belongs. In the attempt which the scholastics made to harmonize the Jehovah of the Jewish culture with the rational God of Aristotle, emphasis was laid on the fact that there is no logical contradiction whatsoever in the idea of the Creator interposing his hand suddenly within the order of His own creation, and thus we have belief in miracles and wonders established on a philosophical basis. Therefore in the philosophy of the historic rationalist school, though the order of nature is admitted as inevitably predetermined by the Supreme Cause, yet the causal chain in the world itself may at any time be interrupted by the intervention of a supernatural power.

We pass now from the Greco-scholastic to the modern philosophical concept of the world. Réné Descartes is generally considered to be the father of modern philosophy. According to Descartes, God made all the laws of nature and all the laws that govern the human spirit by an act of His own free will and for purposes which are so recondite that human thought is unable to penetrate to their full meaning.

Therefore in Cartesian philosophy the possibility of miracles is by no means excluded. Moreover, the logical consequence of the inscrutability of God's design in the world is that we must admit the possibility of events the understanding of which lies entirely outside of the range of the human intellect. These may be called mysteries rather than miracles in the scholastic sense of the latter term. In other words, as our minds are not capable of encompassing the laws which guide the universe we must be content to treat certain happenings as beyond all our power of explanation and referable only to the mysterious ways of Divine Providence. For the purpose of science this means that practically we must admit the existence of breaks in the causal chain.

In contradistinction to the Cartesian Divinity, the God of Baruch Spinoza is a God of harmony and order, whose nature so interpenetrates all creation that the universal causal relation is itself divine and therefore absolutely perfect and permitting of no exceptions. In Spinoza's view of the world there is no room for accident or miracle. That is to say, the causal interrelation is absolutely unbroken.

The next great name that comes into view, when looking over the various world philosophies which were founded on a rationalistic basis, is that of Gottfried Wilhelm Leibniz. According to Leibniz the world was made in fulfilment of a plan corresponding to the supreme wisdom of the Creator. In every created thing God implanted the law of its own individual being, so that each being in the world is independent of and develops independently of all other things, following only the law of its own individual destiny. Therefore, according to Leibniz, the causal inter-relation between one thing and another is only apparent. This means that we must exclude the principle of causation.

We may conclude, I think, from these few examples that the philosophical theories rationally deduced from abstract principles, as regards the place of the causal principle in the world, are almost as numerous as the philosophers themselves. It is obvious that along this road we can make no progress towards a solution of the general problem.

Now we come to a break in the philosophical tradition. Whatever may be said against the English empiricist school and its solipsist[1] consequences, at least it made a break with the naïve conceits of the traditional rationalist school and opened up the way to the development of a philosophical outlook which is more in harmony with the scientific view of the world. The outstanding characteristic in the teaching of the English empirical school is that there is no such thing as certain knowledge or innate ideas, such as were presumed by some of the earlier rationalist philosophers. The human mind as it comes into the world is an absolute blank, on which sense-given impressions are automatically recorded without any action on the part of the mind itself.

John Locke was the founder of this school. He represents the first systematic attempt to estimate in

[1] Solipsism is the theory that the only conscious being is myself.

a critical way the certainty and adequacy of human knowledge when confronted with the universe around it. According to Locke all ideas ultimately depend on experience and by experience Locke means the sensory perceptions of the five senses. Beyond these five senses there is only the reflective consciousness, which is not a sense, as having nothing to do with objects, but as Locke says "it may properly enough be called an internal sense." What we feel to be warm or cold or hard or soft and what we see to be red or blue, that we know; and no other special definition of it is necessary or indeed possible. One often hears of a delusion of the senses, as may happen in the case of a mirage, for instance. This however does not imply that the sensation itself is mistaken, but rather that the conclusions which we draw from the sensory perception are incorrect. What deceives us is not the perceptive sense but the rationalizing intellect.

Sensory perception is something entirely subjective and therefore from this we cannot deduce the existence of the object. Green is not a property of the leaf but a sensation which we experience on looking at the leaf. And so it is with the other senses. Remove the sense-impressions and nothing of the object will remain. John Locke seems to have thought that the sense of touch plays a more important rôle than the other senses, because it is through this sense that we perceive the mechanical qualities of bodies such as thickness, extension, form and movement, and Locke seems to attribute these qualities to something in the bodies themselves. But the later empiricists, especially

David Hume, held that all mechanical qualities of bodies existed only in the senses of the perceiving subject.

In the light of this theory the so-called outer world resolves itself into a complex of sense-impressions and the principle of causation signifies nothing more than a certain order experienced in the sequence of one sensation after another. The idea of order is itself a sense-impression which must be taken as something immediately given and which does not permit of further analysis, for that order may come to an end at any moment. Therefore there is no causation. One thing is observed to follow another but observation cannot assert that it is "caused" by that other thing.

If a rapidly moving billiard-ball strikes against another and sets the latter in motion we experience two independent sense-impressions, one after the other: namely the sensory perception of the moving billiard-ball and the sensory perception of the one set in motion by it. If we stand beside the billiard-table as the play goes on these observations are repeated and we can register a certain regularity between the impressions. For instance, we can perceive that the velocity of the second billiard-ball depends upon the velocity and mass of the billiard-ball that strikes it. We can discover also a further order between these two phenomena. We can, for instance, measure the noise of the impact by its force and we can detect the momentary flattening on each ball at the point of contact with the other ball if we smear one of the balls with some coloured material. All these however are only so many sense-perceptions which accompany one another regularly or displace one another regularly. But they are such that there is no logical connecting-link between the one and the other. If we speak of the *force* which the moving billiard-ball exercises on the one that is at rest, this is only an analogy concept which arises through the muscular sensation which we feel if we ourselves move the ball that is at rest with the naked hand rather than through the medium of the moving billiard-ball. The concept of force has been very useful for the formulation of the laws of motion, but from the viewpoint of knowledge it helps nothing whatsoever. And this is because we have no way of joining up, through a causal bond or a logical bridge, the different phenomena of motion that we have experienced. The individual sense-impressions are different and will remain different, no matter what relations between them may be perceived.

Here the meaning of the principle of causation, taken fundamentally, lies simply in the statement that from the same or similar sensory complexes as cause the same or similar sensory complexes will follow as an effect; but herein the question as to what may be looked upon as similar will on each occasion demand special proof. Formulated in this way, the principle of causation is deprived of all deeper meaning. But this of course does not mean that the law of causation has no practical significance for the human reason. All it means is that the postulate of causation does not furnish us with the grounds of any certain knowledge.

How then can the fact be explained that in common everyday life we take the causal relation of things as something objective and independent? How can this be if in reality we experience nothing more than orderly succession of individual sense-perceptions? The teaching of empirical scepticism answers that this happens through the enormous utility of the causal concept and through the force of habit. Habit certainly plays an important part in life. From childhood onwards it influences our temperament, our wills and our thought. We think we understand a thing merely because we have become accustomed to looking at it. The first time that something new strikes us we feel surprised; but if the same thing happens for the tenth time we find it quite a natural happening. If it should happen a hundred times we say that it is obvious and we even go the length of looking upon it as a matter of necessity. Over one hundred years ago or so mankind in general was acquainted with no other locomotive force except the muscular force of man and beast. As a consequence, no other form of force was considered possible. The pressure of the air and falling water was recognized and applied to mechanical purposes. But here the force itself was stationary and not locomotive in the arbitrary sense. Only men and animals by their muscular effort could move at will from one place to another. A story is told that when the first railways were seen running through the countryside the peasants betted with one another as to how many horses were concealed in the engine. With steam and electric motors everywhere our youth of to-day cannot easily understand the mentality of the peasant of one hundred years ago who felt the necessity of attributing locomotive transport exclusively to a natural horse power.

So far the sceptics are right in saying that it is by force of habit and custom that we attribute certain happenings to certain causes. But at the same time this force of habit cannot explain why we should make the attribution at all. In Fritz Reuter's story *Rei's Nah Belligen*, the peasants undoubtedly made a ludicrous mistake in supposing that there were horses concealed in the steam engine, just as the ancient Greek peasant made a mistake in attributing the thunder to the personal anger of Zeus. But this is not the point here. The point rather is to answer the question why these events should be attributed to a cause at all and how it is that the concept of causation itself arises when we see one event following another. The mere regular succession of impressions does not explain this.

If we go a little deeper into the consideration of the empiricist theory and ask where it would finally lead us were we to pursue it to its logical consequences we shall thus be putting it to a practical test. In the first place we must bear in mind the fact that when there is question of sensory perception as the sole and exclusive source of knowledge, then there can be question only of each one's personal sensory perception in each one's own consciousness. That other men have similar perceptions we can assume only by analogy; but, on the empiricist theory, we cannot know this nor can we logically prove it. Therefore if we are to abide by the logical consequences of the empirical doctrine and exclude all arbitrary assumption, we must confine ourselves, each one of us, to the grounds of his or her own personal sense-perceptions. Then the principle of causation is only a framework for our experiences, connecting them with one another as they enter through the senses and, being entirely unable to tell us anything of what is to come next, it cannot tell us whether the sequence of our experiences may not be broken in a moment. This condition of affairs would seem to obliterate every line of distinction between the sensory perceptions arising from the world of ordinary happenings and those that have no foundation whatsoever in that world. Take the case of sleep for instance. I may dream all sorts of things during the night; but the moment I wake up the reality of my surroundings gives the lie to the dream. The empiricist however cannot logically admit that. For him there is no waking reality; because the subjective sensation is the sole source of awareness in consciousness and is the sole basis and criterion of knowledge. Now the dreamer during the dream believes automatically in its reality and, according to the empiricists, the wideawake person believes automatically in the reality of his sense-perceptions; but has no more reason than the dreamer has for saying that one set of perceptions is false and the other true.

On the grounds of pure logic of course this system of thought, which is commonly called solipsism, is impregnable. The solipsist establishes his ego at the centre of creation, and he does not consider any knowledge as real or sound except that which he for the moment is receiving through his sensory perception. Everything else is derivative and secondary. When the solipsist goes to sleep at night the world ceases to exist for him the moment his eyes and ears and sense of smell and touch become inactive. On rising in the morning everything is new to him again. Here of course I am only imagining what a human being would be if he were a logical consequence of the empirical teaching.

All this of course amounts to a repudiation of common sense; so much so that even the most advanced sceptics of this school find themselves constantly compromising between the claims of common sense and the purely logical conclusions of their own philosophic system. In this connection it is interesting to call attention for a moment to the figure of one of the most outstanding personalities in the subjectivist school, namely Bishop Berkeley. As a student Berkeley studied Locke. But he was of a very deep religious nature and launched a strong criticism against Locke's philosophy because of its scepticism. For Berkeley all things exist only in the mind and the external world can be accounted for only by saying that it exists in the mind of God. He arrives at the existence of God in this way: There are in our own consciousness impressions which are independent of our own wills and sometimes exist even contrary to our wishes. For these impressions we must seek a cause elsewhere than in ourselves and so Berkeley is led to establish the existence of God by practically the same line of reasoning as the rationalist school. For him, however, mind and mind alone exists—the Divine Mind and the human mind. The world of reality as we perceive it exists only in our own mind. Therefore with Berkeley we have no right to talk about a causal interrelation between things in the outer world of reality.

To sum up, empiricism is unassailable on the fundamental ground of a pure logic; and its conclusions are equally impregnable. But if we look at it purely from the viewpoint of knowledge it leads into a blind alley, which is called solipsism. In order to escape from this *impasse* there is no other way open but to jump the wall at some part of it, and preferably at the beginning. This can be done only by introducing, once and for all, a metaphysical hypothesis which has nothing to do with the immediate experience of sense-perceptions or the conclusions logically drawn from them.

Immanuel Kant, the founder of the critical school, was the first to recognize this truth clearly and to point out the way in which the metaphysical step must be taken. According to Kant, the sense-impressions in our consciousness are not the only source of knowledge. The mind has certain concepts that are independent of all experience. These are the so-called categories; and in the philosophy of Kant they are a necessary condition of all knowledge. Kant concluded that causality is such a category. It is one of the ultimate *a priori* forms in which the understanding spontaneously orders its experience—something that is not a derivative from experience but on the contrary is necessary to make orderly experience itself possible. Kant formulated the principle of causality in this way: "Everything that happens presupposes something from which it follows according to a law." Kant held that this postulate is independent of all experience. But Kant's proposition cannot be stated by saying that everything which regularly follows something else has a causal relation to that thing. For instance, there scarcely can be a more regular succession than that of night following day; but nobody would assert that the day is the cause of the night. Succession therefore is not of itself, as with the empiricists, the same as a causal relation. In the example given, namely that of day and night, we have two effects which follow from the same cause. This cause is two-fold. It consists on the one hand of the earth's rotation on its axis and, on the other hand, of the fact that the earth is opaque to the sun's rays.

In the Kantian system therefore the universal validity of the principle of causation is asserted. At the same time, however, it cannot be denied that Kant's teaching, though useful and conclusive in most of its results, is to a certain extent arbitrary on account of its strong dogmatic attitude. This is the reason why it became the subject of so much direct attack and has been altered somewhat with the course of time.

We need not trouble ourselves here with a detailed description of the development of the philosophical side of the causal problem since the time of Kant. It will be sufficient to point out the main features of this development. The strongest opposition to the Kantian doctrine came from the side of those philosophers who maintained that it went too far into the metaphysical field. Now it is perfectly true of course that we cannot avoid metaphysics if we are to save ourselves from falling into the deadlock of solipsism; but, on the other hand, in so far as any system attempts to avoid the metaphysical extreme on the one side and the solipsist extreme on the other, it must be somewhat in the nature of a compromise with logic and therefore will present certain weak features. It is quite possible, however, to construct a system on this basis of compromise wherein the weaker features can be sufficiently strengthened for all practical purposes.

Kant's teaching, and with it the whole of transcendental philosophy from idealism to extreme materialism, is from the outset based on admittedly metaphysical grounds. In contradistinction to this, the positivist system, founded by Auguste Comte, has maintained itself as free as possible in its various shapes and forms from metaphysical influences. It achieves this end by making the experience of our own consciousness the only legitimate source of knowledge. According to the positivist teaching, causality is not founded in the nature of things themselves but is, to put it briefly, an experience of the human mind. It plays an important rôle principally because it has proved itself fruitful and useful. Thus the law of causality is the application of this experience. Because we can always exactly know what we ourselves have discovered by our own experience, the meaning of the causal concept is quite clear to us. But at the same time the possibility remains that there may be cases to which our discovery is not applicable and which therefore contradict the law of causation. Whereas Kant teaches that knowledge without causality is impossible from the very outset, because the category of the causal concept was already in the human mind previous to any experience, the positivist standpoint is that the creative mind of man has fashioned the causal concept for its own convenience. Therefore it is not a primal, inborn quality in the mind. "Man is the measure of all things," said Protagoras long ago. We can twist and turn as we will but we can never get out of our own skins. And whatever tangent we may fly off at into the realm of the absolute we are always really moving around within our own orbit, which has been prescribed for us by the range of experience perceived in our own consciousness. To a certain extent it is not possible to gainsay this positivist attitude, though from the standpoint of transcendental philosophy there are many objections to it. And so argument and counter-argument follow one another in an endless interchange. For us the *dénouement* of the story is the confirmation of our previous conviction, namely, that the nature and universal validity of the Law of Causation cannot be definitely decided upon any grounds of purely abstract reasoning. The transcendental and positivist viewpoints are irreconcilable and they will remain so as long as the race of philosophers lasts.

If pure reasoning had the last word in dealing with such cases then the outlook would be hopeless for any satisfactory settlement of the causative problem. But philosophy, after all, is only one branch of human activity in the study of problems effecting nature and mankind. Science is another branch. And where philosophy has failed in a given instance we are perfectly justified in turning to science and asking whether it may not have a satisfactory answer to suggest.

Now, let us first ask whether the various branches of science are divided against one another on this question of causation, just as philosophy is divided? At the very threshold of this inquiry it may be objected that a problem which falls within the scope of philosophy and which philosophy fails to solve cannot possibly be solved within the limits of a single science. This objection is urged on the grounds that philosophy furnishes the mental foundations on which scientific investigation rests. Philosophy must precede every special science and we should be going against the grain of our whole mental discipline if one of the special sciences were to take up the treatment of general philosophic questions.

That argument is very often urged. But in my opinion the weakness of it is that it leaves out of consideration the collaboration which actually exists between philosophy and the various special sciences. We must remember that the starting-point of all investigation and the mental equipment used in the pursuit of it are fundamentally the same in the case of philosophy as in the case of science. The philosopher does not operate with a kind of human understanding that is special to himself. The structure of thought which he builds up is not based on any other foundation except that of his daily experience and the opinions which he has formed during the course of his professional studies. These latter must largely correspond to his individual talents and the background of his personal philosophical development. In a certain sense the philosopher is in a much higher position than the scientific specialist, because the latter confines observation and research to a much narrower range of facts that are systematically assembled and call for a deep and concentrated kind of probing. Therefore the philosopher has a better outlook on general relations which do not immediately interest the scientific specialist and which may easily pass unobserved by the latter.

The difference between the outlook and work of these two types of investigation may be compared to the case of two travellers who visit the same district together. The first traveller, let us say, is interested in the general features of the landscape, the undulations of hill and valley, and the varying patterns of forest and meadowland. The second traveller is interested only in the flora and fauna or possibly only in the mineral products of the region. His eyes are watching for particular specimens of the former, or he may select various patches of ground for scientific examination in the hope of discovering the presence of mineral wealth beneath. Now the first traveller certainly acquires a better knowledge of the landscape as a whole and can contrast it with other landscapes. From a general view he may conclude in a general way as to the mineral qualities of the soil and the kind of vegetation or animal life that characterize it; but his deductions would be quite general and will depend for verification and clarity of statement on the opinion supplied to him by his companion. Therefore the work of the one is complementary to the work of the other; and there may be innumerable instances wherein the work of the second traveller will be absolutely necessary to the solution of problems which have baffled the man with the more general outlook.

This comparison, like every other comparison, is not fully adequate to the situation. But at least it brings out this point, namely, that in the case of a definite problem which philosophy recognizes as fundamental and the final solution of which is the business of philosophy alone, where philosophy cannot come to a decisive formulation by the use of its own methods it must seek information from the special branches of science in regard to particular features of the problem at issue. Now if the answer here turned out to be definite and final then it must be treated as such. It is a characteristic mark of every true science that the general and objective knowledge which it arrives at has a universal validity. Therefore the definite results which it obtains demand an unqualified acknowledgment and must always hold good. The progressive discoveries of science are definite and cannot permanently be ignored.

This is shown very clearly in the development of natural science. By means of wireless telegraphy we can now send whatever news we wish to the most distant parts of the earth within the infinitesimal fraction of a second. Modern man can lift himself into the air in an aeroplane and transport himself from one part of the globe to the other, over valley and mountain and lake and ocean. By means of X-rays he can pry into the secret activities and inner functions of living organisms and can discover the location of individual atoms in the crystal. This objective achievement which science has accomplished, in collaboration with the technique that it has fertilized, has thrown into the shade some of the greatest discoveries of the philosophers of past times and made a laughing-stock of the crude arts of the magician.

Were anybody to close his eyes to such tangible results and talk about the collapse of science people in general would not think of taking the trouble to refute him. There is no need whatsoever to bring forward any elaborate proof of the contribution to the advancement of knowledge which science has to its credit. It is sufficient merely to point to the events that are before everybody's eyes. One has only to look up when sitting in one's garden and call attention to the drone of the aeroplane or to turn on a switch in one's study and bid the sceptic listen to voices that are coming from a distance of thousands of miles. The worth of any human endeavour is and always must be the results which it has obtained.

Now let us return to the particular problem that we are dealing with and let us admit for the moment the competence and reliability of the scientific method in the treatment of it. Let us ask how does science, in each of its different branches, actually regard the problem of causation. Here it must be remembered that I am talking of specialized science as such and not of the philosophical or epistemological foundations on which it works. Does science as a matter of fact occupy itself exclusively with data immediately given by sensory impressions and their systematic organization according to laws of reason? Or does it at the very outset of its activities reach out beyond the knowledge given us by this immediate source and make, as it were, a jump into the metaphysical sphere?

I do not think that there can be any doubt whatsoever as to the answer. The first alternative is ruled out and the second affirmed in the case of each special science. Indeed it may be said that every individual science sets about its task by the explicit renunciation of the ego-centric and anthropocentric standpoint. In the earlier stages of human thought mankind turned its attention exclusively to the impressions received through the senses, and primitive man made himself and his own interests the centre of his system of reasoning. Confronted with the powers of nature around him, he thought that they were animated beings like himself and he divided them into two classes, the one friendly and the other inimical. He divided the plant world into the categories of poisonous and non-poisonous. He divided the animal world into the categories of dangerous and harmless. As long as he remained bound within the limits of this method of treating his environment it was impossible for him to make any approach towards real scientific knowledge. His first advance in this knowledge was accomplished only after he had taken leave of his own immediate interests and banished them from his thought. At a later stage he succeeded in abandoning the idea that the planet whereon he lives is the central point of the universe. Then he took up the more modest position of keeping as far as possible in the background, so as not to intrude his own idiosyncrasies and personal ideas between himself and his observations of natural phenomena. It was only at this stage that the outer world of nature began to unveil its mystery to him, and at the same time to furnish him with means which he was able to press into his own service and which he could never have discovered if he had continued looking for them with the candlelight of his own egocentric interests. The progress of science is an excellent illustration of the truth of the paradox that man must lose his soul before he can find it. The forces of nature, such as electricity for instance, were not discovered by men who started out with the set purpose of adapting them for utilitarian purposes. Scientific discovery and scientific knowledge have been achieved only by those who have gone in pursuit of it without any practical purpose whatsoever in view. The few examples that I have mentioned make this abundantly clear. Heinrich Hertz, for instance, never dreamt that his discoveries would have been developed by Marconi and finally evolved into a system of wireless telegraphy. And Roentgen could never have called up a vision of the immense range of beneficial purposes to which the X-rays are applied to-day.

I have said that the first step which every specialized branch of science takes consists of a jump into the region of metaphysics. In taking this jump the scientist has confidence in the supporting quality of the ground whereon he lands, though no system of abstract reasoning could have previously assured him of that. In other words, the fundamental principles and indispensable postulates of every genuinely productive science are not based on pure logic but rather on the metaphysical hypothesis—which no rules of logic can refute—that there exists an outer world which is entirely independent of ourselves. It is only through the immediate dictate of our consciousness that we know that this world exists. And that consciousness may to a certain degree be called a special sense. And one may go even so far as to say that the existence of the exterior world strikes the consciousness of each individual in some particular way. It is as if we looked at some distant object through a pair of glasses and as if each one were wearing glasses of a slightly different shade of colour. And we must take this into account when we deal scientifically with natural phenomena. The first and most important quality of all scientific ways of thinking must be the clear distinction between the outer object of observation and the subjective nature of the observer.

Once the scientist has begun by taking his leap into the transcendental he never discusses the leap itself nor worries about it. If he did science could not advance so rapidly. And anyhow—which is fundamentally a consideration of no less importance—this line of conduct cannot be refuted as inconsistent on any logical grounds.

Of course there is the positivist theory that man is the measure of all things. And that theory is irrefutable in so far as nobody can object on logical grounds to the action of a person who measures all things with a human rule, and resolves the whole of creation ultimately into a complex of sensory perceptions. But there is another measure also, which is more important for certain problems and which is independent of the particular method and nature of the measuring intellect. This measure is identical with the *thing* itself. Of course it is not an immediate datum of perception. But science sets out confidently on the endeavour finally to know the *thing* in itself, and even though we realize that this ideal goal can never be completely reached, still we struggle on towards it untiringly. And we know that at every step of the way each effort will be richly rewarded. The history of science is at hand to confirm our faith in this truth.

Having once assumed the existence of an independent external world, science concomitantly assumes the principle of causality as a concept entirely independent of sense-perception. In applying this principle to the study of natural phenomena science first investigates if and how far the law of causal relation is applicable to the various happenings in the world of nature and in the realm of the human spirit. Science finds itself here exactly on the same footing which Kant took as the starting-point of his theory of knowledge. As in the case of Kantian philosophy, so also in the case of each special branch of science the causal concept is accepted at the outset as belonging to those categories without which no progress in knowledge can be made. But we must make a certain differentiation here. Kant took not merely the concept of causality but also to a certain degree the meaning of the causal law itself as an immediate datum of knowledge and therefore universally valid. Specialized science cannot go thus far. It must rather confine itself to the question as to what significance the law of causality can be proved to have in each individual case, and thus through research give practical meaning and value to the empty framework of the causal concept.

CHAPTER V

CAUSATION AND FREE WILL

THE ANSWER OF SCIENCE

WE now come to ask whether and how far science can help us out of the obscure wood wherein philosophy has lost its way. What is the practical attitude adopted by the special sciences in regard to the universal and invariable validity of the law of causation? Does science in its everyday investigations accept the principle of causation as an indispensable postulate? Does it act upon the assumption that there are no loopholes in the causally governed order of nature? Or, while using the principle as a working hypothesis, does scientific practice intimate that there are certain happenings in nature where the law of causation does not function, and that there are regions in the mental sphere where the causal writ does not run? In our endeavour to find a definite answer to those questions we shall have to put them singly to each of the several branches of specialized science. In doing this of course we shall have to be content with quite a summary cross-examination. What has physical science to say to our problem? What has the science of biology to answer? And what have the humanist sciences, such as psychology and history, to say?

Let us begin with the most exact of the natural sciences, namely, physics. In classical dynamics, among which we must include not only mechanics and the theory of gravitation but also the Maxwell-Lorentz view of electrodynamics, the law of causality has been given a formulation which for exactitude and strictness may be considered almost as ideal, even though it may be somewhat one-sided. It is expressed in a system of mathematical equations through which all happenings in any given physical picture can be absolutely predicted if the time and space conditions are known—that is to say, if the initial state be known and the influences which are brought to bear upon the picture from outside. To put the matter in a more concrete way: according to the law of causation as expressed in the equations of classical dynamics, we can tell where a moving particle or system of particles may be located at any given future moment if we know their location and velocity now and the conditions under which the motion takes place. In this way it was made possible for classical dynamics to reckon beforehand all natural processes in their individual behaviour and thus to predict the effect from the cause. The last significant advance which classical dynamics achieved in our day came about through the general relativity theory of Einstein. This theory welded together Newtonian gravitation and Galileo's law of inertia. Several attempts have been made recently to show that the relativity theory corroborates the positivist attitude and in a certain sense is incompatible with transcendental philosophy. These attempts are entirely mistaken. For the foundation of the relativity theory is not based on the rule that all time and space dimensions have only a relative meaning, which is determined by the reference system of the observer. The foundation of the relativity theory lies in the fact that in the four-dimensional spacetime manifold there is a measure, namely the distance between two points approximating with infinite closeness. This is the so-called Tensor or *Massbestimmung*, which for all measuring observers and for all reference systems has the self-same value, and it therefore is of a transcendental character entirely independent of any arbitrary action of the human will.

Into this harmonized system of classical-relativist physics, however, the quantum hypothesis has recently introduced a certain disturbance, and one cannot yet definitely say what influence the subsequent development of the hypothesis may have on the formulation of fundamental physical laws. Some essential modification seems to be inevitable; but I firmly believe, in company with most physicists, that the quantum hypothesis will eventually find its exact expression in certain equations which will be a more exact formulation of the law of causality.

Besides dynamical laws applied to individual cases, physical science recognizes other laws also, which are called statistical. These latter express to a fairly accurate degree the probability of certain happenings occurring and therefore they allow for exceptions in particular cases. A classical example of this is the conduction of heat. If two bodies of different temperatures be brought into contact with one another then, according to the two laws of thermodynamics, the heat energy will always pass from the warmer to the cooler body. We know to-day from experiment that this law is only a probability; because, especially when the difference of temperatures between two bodies is exceptionally small, it may well happen that at one or other particular point of contact and at one particular moment of time the conduction of heat will take place in the opposite direction—that is to say, from the cooler to the warmer body. The second law of thermodynamical, as in the case of all statistical laws, has an exact significance only for average values arising from a great number of similar happenings and not for each happening itself. If we are to consider the individual happening we can speak only of a definite measure of probability. The case here is quite similar to the case of a non-symmetrical cube used in playing with dice. Let us suppose that the centre of gravity of the cube is not at the centre of the body but lies definitely towards one of the sides, then it is likely though by no means certain that when the cube is thrown it will come to rest on that side. The smaller the distance of the centre of gravity from the symmetrical centre of the cube the more variable will the result be. Now if we cast the dice sufficiently often and observe what happens in each case, then we can arrive at a law which will tell us that the dice will fall on a certain side so many times out of a thousand for instance.

Let us return to the example of heat conduction and ask whether the strict validity of the causal law holds for individual cases. The answer is that it does hold; because more thoroughgoing methods of investigation have proved that what we call transfer of heat from one body to another is a very intricate process, unfolding itself through innumerable series of particular processes which are independent of one another and which we call molecular movements. And investigation has further shown that if we presuppose the validity of dynamical laws for each of these particular happenings—that is to say, the law of strict causality—then we can arrive at the causal results through this type of observation. In point of fact, statistical laws are dependent upon the assumption of the strict law of causality functioning in each particular case. And the non-fulfillment of the statistical rule in particular cases is not therefore due to the fact that the law of causality is not fulfilled, but rather to the fact that our observations are not sufficiently delicate and accurate to put the law of causality to a direct test in each case. If it were possible for us to follow the movement of each individual molecule in this very intricate labyrinth of processes, then we should find in each case an exact fulfillment of the dynamical laws.

In speaking of physical science under this aspect we must always distinguish between two different methods of research. One is the macroscopic method, which deals with the object of research in a general and summary manner. The other is the microscopic method, which is more delicate and detailed in its procedure. It is only for the macroscopic observer— that is to say, the man who deals with big quantities in a wholesale way—that chance and probability exist in regard to single elements in the object that he handles. The extent and importance of the chance elements are of course dependent on the measure of knowledge and skill which is brought to bear on the object. On the other hand, for the microscopic investigator only accuracy and strict causality exist. His livelihood depends, as it were, on the quality of each individual item that he deals with in retail. The macroscopic investigator reckons only with mass values and knows only statistical laws. The microscopic investigator reckons with individual values and applies to them dynamical law in its full significance.

Suppose we consider again the example of the dice which I have mentioned already. And suppose we treat it microscopically. This means that together with the nature of the dice itself—its non-symmetrical character and the exact location of its centre of gravity —we also take into account its initial position and its initial velocity and the influence of the table on its movement, the resistance of the air and every other peculiarity that may affect it—supposing we could examine all these minutely, then there could be no question of chance; because each time we can reckon the place where the dice would stop and know in what position it would rest.

Without going into any further details, let me say that physical science applies the macroscopic method of research to all happenings where molecules and atoms are concerned. But it naturally strives to refine its treatment towards the microscopic degree of delicacy and always seeks to reduce statistical laws to a dynamic and strictly causal system. Therefore it may be said here that physical science, together with astronomy and chemistry and minerology, are all based on the strict and universal validity of the principle of causality. In a word, this is the answer which physical science has to give to the question asked at the beginning of the present chapter.

Let us come now to the science of biology. Here the conditions are very much more intricate, because biology deals with living things and the problem of life has always presented very serious difficulties for scientific research. Of course I cannot speak with special authority in this branch of science. Yet I have no hesitation in saying that even in the most obscure problems, such as the problem of heredity, biology is approaching more and more to the explicit assumption of the universal validity of causal relations. Just as no physicist will in the last resort acknowledge the play of chance in inanimate nature, so no physiologist will admit the play of chance in the absolute sense, although of course the microscopic method of research is very much more difficult to carry out in physiology than in physics. For this latter reason the majority of physiological laws are of a statistical character and are called rules. When an exception occurs in the application of these empirically established rules, this is not attributed to any skip or failure in the causal relation but rather to a want of knowledge and skill in the way that the rule is applied. The science of biology sets its face against permitting exceptions as such to exist. What appear to be exceptions are carefully recorded and collated and are further studied until they are cleared up in the light of causal relations. Very often it happens that this further study of exceptions shows interrelations which were hitherto unthought of, and throws a new light on the rules under which the exceptions were originally found to occur. It very often happens that the universal causal relation is thus corroborated from a new side, and that is the way in which many significant discoveries have been made.

How can we distinguish between what is veritably a causal relation and what is merely a coincidence or external succession of one event following another? The answer is that there is no hard and fast rule for making such a distinction. Science can only accept the universal validity of the law of causation, which enables us definitely to predict effects following a given cause, and in case the predicted effect should not follow then we know that some other facts have come into play which were left out of consideration in our reckoning. A little story will illustrate my meaning here. It refers to the efficiency of artificial manure in agriculture.

If I am not mistaken the story is told of Benjamin Franklin. He was not merely a first-class statesman but he was also a very able research worker and discoverer in natural science. At one time he took a great interest in the problem of artificial manuring and clearly showed the importance of its development in agrarian economics. He put his theories to the test and achieved practical successes which were quite satisfying for his own scientific bent of mind. But he

found it very difficult to convince his sceptical neigh-
bours that the luxuriant crop of clover which they saw
growing in Franklin's field was due to the use of
artificial manure. For the peasant clover was clover
and land was land and there were good land and bad
land and good weather conditions and bad weather
conditions, and these were the only factors that he
recognized as causes of a good crop or a poor crop.
Franklin determined to convince the peasant that the
art of man could directly influence the quality of
nature's growth. At the time of seed-sowing he dug
in the soil a series of small furrows which formed
alphabetical letters. These small furrows he filled
with rich quantities of artificial manure, while the other
parts of the field were left solely to nature's hand. As
the crop grew the letters that corresponded to the
manured furrows showed rows of clover much taller
and more luxuriant than that in the other parts of
the field; so that the passers-by were able to read the
sentence: *This part has been manured with gypsum.*
History does not relate whether the obstinate peasants
were or were not convinced by the proof. But that
is neither here nor there; for nobody can be forced
on purely logical grounds to acknowledge the causal
connection, because the causal connection is not logically demonstrable. The point of the illustration here
is that if in a particular case we introduce a cause
which of its very nature "flows into" the result, as the
scholastics used to say, and if the result is in full
accord with what was predicted, then we can be
certain of the causal relation. In the instance of

Franklin's clover there could possibly be no other explanation except that of the manuring, and this explanation, as a cause, has a natural and exclusive connection with the result.

Of course it may be said that the law of causality is only after all an hypothesis. If it be an hypothesis it is not an hypothesis like most of the others, but it is a fundamental hypothesis because it is the postulate which is necessary to give sense and meaning to the application of all hypotheses in scientific research. This is because any hypothesis which indicates a definite rule presupposes the validity of the principle of causation.

We now come to those sciences which deal with human events. Here the method which the scientist follows can have nothing like the same exactitude as that which he follows in physics. The object of his study is the human mind and its influence on the course of events. The great difficulty here is the meagre supply of source materials. While the historian or the sociologist strives to apply purely objective methods to his lines of investigation, he finds himself confronted on all hands with the want of data whereby he might determine the causes that have led to general conditions in the past and lead to the general conditions in the world at the present moment. At the same time however he has at least one advantage here which the physicist has not. The historian or the sociologist is dealing with the same kind of activities as he finds in himself. Subjective observation of his own human nature furnishes him with at least a rough means of estimation in dealing with outside personalities or groups of personalities. He can "feel into" them as it were and may thus gain a certain insight into the characteristics of their motives and their thoughts.

Let us ask then what is the attitude of the humanist scientist towards this problem of causation. In the activities of the human mind and in the play of human emotions, and in the outer conduct that results from these, is there everywhere a rigid causal interrelation? And is all conduct in the last resort to be attributed to the causal activity of circumstances, such as past events and present surroundings, leaving no place whatsoever for an absolutely spontaneous action of the human will? Or have we here, in contradistinction to nature, at least a certain degree of freedom or arbitrary volition or chance, whichever name one wishes to choose? From time immemorial this question has been a source of controversy. Those who hold that the human will is absolutely free in its act of volition generally assert that the higher we go in the scale of natural being the less noticeable is the play of necessity and the greater the play of creative freedom, until we finally come to the case of human beings, who enjoy the full autonomy of the will.

Such an opinion cannot be spoken of as correct or incorrect except by putting it to the test of historical and psychological research. And here we have the problem in exactly the same position as in the case of physical science. In other words we cannot know how far the principle of causality is valid except by putting

it to the test of outer reality. Of course a different
terminology is used when causal methods are applied
in the humanist sciences. In natural science a definite
physical picture with given characteristics is the
subject of research. In psychology we have a definite
individual personality to study. That individual personality has inherited qualities such as bodily con-
formation, intelligence, imaginative capacity, temperament, personal tastes and so on. Working on
this personality we have the physical and psychic
influences of the environment, such as climate, food,
upbringing, companionship, family life, education,
reading, etc. Now the question is whether all these
data determine the conduct of this personality in all
its particulars and according to definite laws. In other
words if we suppose, what is impossible in practice,
that we had a thorough and detailed knowledge of all
these factors here and now, could we tell with certainty,
on the causal basis, how the individual will act a
moment hence?

In seeking for a sound and logical and adequate
answer to this question we are here in quite a different
position from that in which we were when dealing
with natural science. Obviously it is extremely difficult to give anything like a definite answer to such a
question as that asked above. One may have opinions
and make suppositions and assumptions; but these
do not furnish logical grounds for an answer. Still I
think that it may be said definitely that the direction
in which the humanist sciences, such as psychology and
history, are developing nowadays furnishes certain

grounds for presuming that the question should be answered in the affirmative. The part which force plays in nature, as the cause of motion, has its counterpart in the mental sphere in motive as the cause of conduct. Just as at each and every moment the motion of a material body results necessarily from the combined action of many forces, so human conduct results with the same necessity from the interplay of mutually reinforced or contradicting motives, which partly in the conscious and partially also in the unconscious sphere work their way forward towards the result.

Of course it is perfectly true that many acts which are done by human beings appear to be inexplicable. At times it is an extraordinarily difficult riddle to find anything like reasonable grounds for certain acts, and other acts seem so utterly foolish as to suggest no grounds at all. But consider for a moment the way these acts appear to a trained psychologist and the way they appear to the ordinary man in the street. What is entirely puzzling to the latter is often quite clear to the former. Therefore if we could study the acts of the human being at very close and intimate quarters, we should find that they can be accounted for through causes which lie in the character or in the momentary emotional tension or in the specific external environment. And in those cases where it is extremely difficult and wellnigh impossible to discover these explanatory causes, then we have at least grounds for assuming that if we cannot find any motive as an explanation, we must attribute this not actually to the absence of motive but rather to the unsatisfactory nature of our knowledge of the peculiarities of the situation. Here we have the same case as in the throwing of the unsymmetrical dice. We know that the way in which the dice finally comes to rest is the nett result of all the factors active in the throwing of the dice, but in the case of a single throw we cannot detect the function of strict causality. And so, even though the motive of a certain line of human conduct may often lie utterly hidden, conduct entirely without motive is scientifically just as incompatible with the principles on which mental science is carried on as the assumption of absolute chance in inorganic nature is incompatible with the working principle of physical science.

It is not merely however that conduct is conditioned by the motives which lead to it. Each act has also a causal influence on subsequent behaviour. And so in the interchange of motive and conduct we have an endless chain of events following one another in the spiritual life, in which every link is bound by a strict causal relation not only with the preceding link but also with the following one.

Attempts have been made to find a way to free these links from the causal chain. Hermann Lotze, in open contradiction to Kant, put forward the suggestion that such a causal chain can have no end, although it has a beginning. In other words, that circumstances occur in which motives appear entirely independently, not caused by any preceding influence, so that the conduct to which these motives lead will be the first link in a new chain. Such an interpretation, Lotze held, must be given especially to the acts of those choice spirits that are called creative geniuses.

Even though we may not question the possibility of such cases happening in the world of reality yet we may reasonably answer that the thoroughgoing scientific research which has been carried on in the region of psychology would have pointed to such a possibility. But as far as psychological research has gone there are no indications which might furnish a starting-ground for this theory of the so-called free beginning. On the contrary, the deeper scientific research goes into the peculiarities that have characterized even the great spiritual movements of world history, more and more the causal relation emerges into the open. The dependence of each event upon preceding fact and preparatory factors gradually begins to appear under the strong light of scientific investigation, so much so as to warrant the statement that present-day scientific procedure in psychology is founded practically exclusively on the principle of causal inter-relations and the assumption of an active law of causality which permits no exceptions. This means that the postulate of complete determinism is accepted as a necessary condition for the progress of psychological research.

Under these circumstances it is obvious that we cannot erect a definite boundary and say: Thus far but no farther. The principle of causality must be held to extend even to the highest achievements of the human soul. We must admit that the mind of each one of our greatest geniuses—Aristotle, Kant or Leonardo, Goethe or Beethoven, Dante or Shakespeare—even at the moment of its highest flights of thought or in the most profound inner workings of the soul, was subject to the causal fiat and was an instrument in the hands of an almighty law which governs the world.

The average reader may be easily taken aback by such a statement. It may sound derogatory to speak thus of the creative achievements of the highest and noblest of the human race. But on the other hand it must be remembered that we ourselves are only common mortals, and that we could never hope to be in a position to follow out the delicate play of cause and circumstance in the soul of the genius. There is nothing derogatory in saying that they are subject to the law of cause and effect, though it would be derogatory of course if this were interpreted in the sense that the ordinary mortal is capable of following the workings of that law in the case of supremely gifted souls. Nobody would feel it disrespectful if one were to say that some superhuman intelligence could understand a Goethe or a Shakespeare. The whole point lies in the inadequacy of the observer. Just so the macroscopic physicist is entirely unable to pursue microscopic workings in natural phenomena, yet, as we have seen, this does not mean that the law of causality is not valid for these microscopic happenings.

Where is the sense then, it may here be asked, in talking of definite causal relations in regard to cases wherein nobody in the world is capable of tracing their function?

The answer to that question is simple enough. As has been said again and again, the concept of causality is something transcendental, which is quite independent of the nature of the researcher, and it would be valid even if there were no perceiving subject at all. We shall see more clearly the inner meaning of the causal concept if we consider the following:—

At this present moment of time and space the human intellect as we know it may possibly not be the highest type of intellect in existence. Higher intelligences may exist in other places or may appear in other epochs. And the intellectual level of these beings may be as much above ours as ours is above the protozoa. Then it may well happen that before the penetrating eye of such intelligences even the most fleeting moment of mortal thought, as well as the most delicate vibration in the ganglia of the human brain, could be followed in each case, and that the creative work of our mortal geniuses could be proved by such an intelligence to be subject to unalterable laws, just as the telescope of the astronomer traces the links of the manifold movement of the spheres.

Here, as everywhere else, we must differentiate between the validity of the causal principle and the practicability of its application. Under all circumstances the law of causation is valid, because of its transcendental character. But as its application can be carried out in full detail only by the microscopic observer in natural science, so in the region of the human mind the law can be applied only by an intelligence that is far superior to the object of research. The smaller the distance between the investigator and the object in this case, the more uncertain and fallible will be the causal and scientific treatment. The whole problem lies in the difficulty, indeed the impossibility, with which we are faced in trying to understand the behaviour of a genius from the standpoint of causation. Even a congenial spirit in such cases would have to be content with presumptions and analogies; but to the average blockhead the genius will ever remain a closed book signed with the seven seals.

The conclusion therefore is that the highest types of human intelligence are subject to the causal law in the processes that result in even their greatest achievements. That is the first part of our conclusion. And the second part is that in principle we must reckon with the possibility that a day will come when the more profound and increasingly more refined development of scientific research will be able to understand the mental workings not only of the ordinary mortal but also of the highest human genius in their causal relations; because scientific thought is identical with causal thought, so much so that the last goal of every science is the full and complete application of the causal principle to the object of research.

From all that I have said what conclusion are we to draw in regard to Free Will? In the midst of a world where the principle of causation prevails universally,

what room is there for the autonomy of human volition? This is an important question, especially to-day,
because of a widespread tendency unwarrantably to
extend the tenets of scientific determinism to human
conduct and thus shelve responsibility from the
shoulders of the individual. We have had an example
of this in some modern interpreters of historical
development who would hold that the destiny of a
group of individuals, forming a nation or a civilization, is determined by blind fate. Therefore in the
last analysis the responsibility for such a destiny does
not rest with the individual. Is this attitude a legitimate
deduction from all that I have said? In other words,
amid the all-round causal sequence in natural phenomena is there still room for the free and responsible
act of the will of the individual?

Before directly answering that question I may
point to a notable characteristic of everyday life which
may help us in forming a decision. Though chance
and miracle in the absolute sense are fundamentally
excluded from science, yet science is confronted
to-day, more than ever before perhaps, with a widespread belief in miracle and magic. Such belief, which
has been so universal in former ages, repeats itself
with the passing of the centuries in innumerable
forms. This means that science is repeatedly called
upon to give the scientific causal explanation of facts
that are popularly interpreted in the light of some
belief. Belief in miracle is a very important element
in the cultural history of the human race. It has brought
untold blessings and has inspired noble men to the

greatest of heroic deeds. But where it has degenerated into fanaticism it has also been the cause of untold evil.

In view of the remarkable progress of physical science during our own time and the universal extension of its benefits amongst civilized nations, we might naturally assume that one of the achievements of science would have been to restrict belief in miracle. But it does not seem to do so. The tendency to believe in the power of mysterious agencies is an outstanding characteristic of our own day. This is shown in the popularity of occultism and spiritualism and their innumerable variants. Though the extraordinary results of science are so obvious that they cannot escape the notice of even the most unobservant man in the street, yet educated as well as uneducated people often turn to the dim region of mystery for light on the ordinary problems of life. One would imagine that they would turn to science, and it is probably true that those who do so are more intensely interested in science and are perhaps greater in number than any corresponding group of people in former times; but still the fact remains that the drawing power of systems which are based on the irrational is at least as strong and as widespread as ever before, if not more so. The Monist League which was formed some years ago with so much *éclat* and promise, for the purpose of establishing a world outlook based on purely scientific grounds, has certainly not achieved any success corresponding to the rival systems.

How is this peculiar fact to be explained? Is there,

in the last analysis, some basically sound foothold
for this belief in miracle, no matter how bizarre and
illogical may be the outer forms it takes? Is there
something in the nature of man, some inner realm,
that science cannot touch? Is it so that when we
approach the inner springs of human action science
cannot have the last word? Or, to speak more con-
cretely, is there a point at which the causal line of
thought ceases and beyond which science cannot go?

This brings us to the kernel of the problem in regard
to free will. And I think that the answer will be found
automatically suggested by the questions which I have
just asked.

The fact is that there is a point, one single point
in the immeasurable world of mind and matter, where
science and therefore every causal method of research
is inapplicable, not only on practical grounds but also
on logical grounds, and will always remain inapplicable.
This point is the individual ego. It is a small point in
the universal realm of being; but in itself it is a whole
world, embracing our emotional life, our will and
our thought. This realm of the ego is at once the
source of our deepest suffering and at the same time
of our highest happiness. Over this realm no outer
power of fate can ever have sway, and we lay aside
our own control and responsibility over ourselves only
with the laying aside of life itself.

And yet there is a way in which the causal method
can be applied within the limits of this inner realm.
In principle there is no reason whatsoever why the
individual should not make himself the observer of

L

what has happened within himself. In other words, he can look back over the experiences through which he has passed and endeavour to link them up in their causal relations. There is no reason indeed, at least in principle, why he should not scrutinize each experience—by which I mean each decision and line of conduct which he has taken—and study it from the viewpoint of finding out the cause from which it resulted. Of course that is an extremely difficult task; but it is the only soundly scientific way of dealing with our own lives. In order to carry out this plan of action the facts of our own lives which we now place under observation would have to be distanced in the past, so that our present complex of living emotions and inclinations would not enter as factors into the observation. If we could possibly carry out the plan in this detached way, then each experience through which we have passed would make us immeasurably more intelligent than we were before, so intelligent indeed that in relation to our earlier condition we should rise to the level of the super-intelligence postulated by Laplace. You remember that Laplace held that if there were a super-intelligence standing entirely outside of the facts occurring in the universe, this intelligence would be able to see causal relations in all the happenings of the world of man and nature, even the most intricate and microscopic. It is only by aiming at this sort of distance that the individual could establish the required detachment of the perceiving subject from the object of his research, which we have already seen to be an inevitable condition for the application of the causal method in research. The nearer we are to events in time the more difficult it is to trace their causal structure. And the nearer we are to the events of our own personal experience the more difficult it is for us to study ourselves in the light of these happenings; for the activities of the observer are here partly the object of research and, in so far as that is so, the causal connection is practically impossible to establish. I am not preaching a moral sermon here or suggesting what ought to be aimed at for the sake of the moral uplift of one's own being. I am only treating the case of individual freedom from the viewpoint of its logical coherence with the principle of causation, and I am saying that *in principle* there is no reason why we should not discover the causal connections in our own personal conduct, but that in practice we never can do so because this would mean that the observing subject would also be the object of research. And that is impossible; for no eye can see itself. But in so far as any man is not entirely to-day that which he was years ago there is a relative degree to which he might subject his own experiences to causal scrutiny; and I have mentioned this as illustrative of the general principle.

It will occur to many readers to ask if thus in relation to the chain of causality the freedom of the individual will, here and now, is only apparent and results solely from the defects of our own understanding. That way of putting the case is, I am convinced, entirely mistaken. We might illustrate the mistake by saying that it is like the mistake of suggesting that the inability of a runner to outrun his own shadow is due to his lack of speed. The fact that the individual here and now, in regard to his own living present act, cannot be subject to the law of causation is a truth that is based on a perfectly sound logical foundation of an *a priori* kind, such as the axiom that the part is never greater than the whole. The impossibility of the individual contemplating his own activity here and now under the light of the causal principle would hold good even in the case of the super-intelligence postulated by Laplace. For, even though this super-intelligence might be able to trace the causal structure in the achievements of the most gifted geniuses of the human race, yet that same super-intelligence would have to renounce the idea of studying the activities of its own ego at the moment it contemplated the activities of our mortal ego. If there be a Supreme Wisdom whose celestial nature is infinitely elevated above ours, and who can see every convolution in our brains and hear every pulse beat of each human heart, as a matter of course such a Supreme Wisdom sees the succession of cause and effect in everything we do. But this does not in the least invalidate our own sense of responsibility for our own actions. From this standpoint we are on an equal footing with the saints and confessors of the most sublime religions. We cannot possibly study ourselves at the moment or within the environment of any given activity. Here is the place where the freedom of the will comes in and establishes itself, without usurping the right of any rival. Being emancipated thus, we are at liberty to construct any miraculous background that we like in the mysterious realm of our own inner being, even though we may be at the same time the strictest scientists in the world, and the strictest upholders of the principle of causal determinism. It is from this autarchy of the ego that the belief in miracles arises, and it is to this source that we are to attribute the widespread belief in irrational explanations of life. The existence of that belief in the face of scientific advance is a proof of the inviolability of the ego by the law of causation in the sense which I have mentioned. I might put the matter in another way and say that the freedom of the ego here and now, and its independence of the causal chain, is a truth that comes from the immediate dictate of the human consciousness.

And what holds good for the present moment of our being holds good also for our own future conduct in which the influences of our present ego play a part. The road to the future always starts in the present. It is, here and now, part and parcel of the ego. And for that reason the individual can never consider his own future purely and exclusively from the causal standpoint. That is the reason why fancy plays such a part in the construction of the future. It is in actual recognition of this profound fact that people have recourse to the palmist and the clairvoyant to satisfy their individual curiosity about their own future. It is also on this fact that dreams and ideals are based, and here the human being finds one of the richest sources of inspiration.

I might mention here in passing that this practical inapplicability of the law of causation extends beyond the individual. It extends to our relations with our fellow-men. We are too much a part of the life of our fellow beings to be in a position to study them from the viewpoint of motives, which means the causal viewpoint. No ordinary human being can put himself in the position of the super-intelligence imagined by Laplace and consider himself capable of tracing all the inner springs of action from which the conduct of his fellow-men originates. On the other hand, however, I would mention here again a phase of the causal application corresponding to that which I have already spoken of in relation to the individual's capacity for scientifically observing his own past experience. To a relative degree it is possible to study the motives on which other people act, just as they are studied by the psychologist or the alienist. In all such cases there is to a certain degree the requisite distance between the researcher and the object of his research. And therefore to this extent there is no logical incoherence in the idea of a person studying the activities of his fellow beings. Indeed all who wish to influence others do so in everyday life, which is largely the secret of political success. It is the secret of all the power for good which so many people exercise in relation to their fellow beings. Most of us remember from childhood personalities whom we shirked because of some sort of innate feeling of

insecurity in their presence, and on the other hand
most of us, I imagine, have memories of acquaintances
to whose influence we were willingly amenable because
we felt a certain reverence towards them. And everybody is more or less familiar with the feeling of
withdrawal which comes over one in the presence of
a person who is suspected of seeing too clearly into
the inner lives of others. All these immediate reactions
bear witness to a sort of instinctive recognition that
our own lives are in the last analysis subject to causation, though the ego as regards its immediate destiny
cannot be subject to that law.

Science thus brings us to the threshold of the ego
and there leaves us to ourselves. Here it resigns us
to the care of other hands. In the conduct of our own
lives the causal principle is of little help; for by the
iron law of logical consistency we are excluded from
laying the causal foundations of our own future or
foreseeing that future as definitely resulting from the
present.

But mankind has need of fundamental postulates
for the conduct of everyday existence, and this need
is far more pressing than the hunger for scientific
knowledge. A single deed often has far more significance for a human being than all the wisdom of the
world put together. And therefore there must be
another source of guidance than mere intellectual
equipment. The law of causation is the guiding rule
of science; but the Categorical Imperative—that is to
say, the dictate of duty—is the guiding rule of life.
Here intelligence has to give place to character, and

scientific knowledge to religious belief. And when I say religious belief here I mean the word in its fundamental sense. And the mention of it brings us to that much discussed question of the relation between science and religion. It is not my place here nor within my competency to deal with that question. Religion belongs to that realm that is inviolable before the law of causation and therefore closed to science. The scientist as such must recognize the value of religion as such, no matter what may be its forms, so long as it does not make the mistake of opposing its own dogmas to the fundamental law upon which scientific research is based, namely the sequence of cause and effect in all external phenomena. In conjunction with the question of the relations between religion and science, I might also say that those forms of religion which have a nihilist attitude to life are out of harmony with the scientific outlook and contradictory to its principles. All denial of life's value for itself and for its own sake is a denial of the world of human thought, and therefore in the last analysis a denial of the true foundation not only of science but also of religion. I think that most scientists would agree to this, and would raise their hands against religious nihilism as destructive of science itself.

There can never be any real opposition between religion and science; for the one is the complement of the other. Every serious and reflective person realizes, I think, that the religious element in his nature must be recognized and cultivated if all the powers of the human soul are to act together in perfect balance and harmony. And indeed it was not by any accident that the greatest thinkers of all ages were also deeply religious souls, even though they made no public show of their religious feeling. It is from the cooperation of the understanding with the will that the finest fruit of philosophy has arisen, namely the ethical fruit. Science enhances the moral values of life, because it furthers a love of truth and reverence —love of truth displaying itself in the constant endeavour to arrive at a more exact knowledge of the world of mind and matter around us, and reverence, because every advance in knowledge brings us face to face with the mystery of our own being.

CHAPTER VI

FROM THE RELATIVE TO THE
ABSOLUTE

I HOPE the reader will not be frightened away by the sound of this title. I should have chosen another terminology if I could have found one better suited to my purpose. But the above title is the most expressive I can find to indicate an outstanding feature of scientific development which I wish to describe here. This feature has been remarkably characteristic of physical science during the past hundred years. The line of progress has been from the relative to the absolute. We need not delay here to discuss the various meanings given to these words in scientific and semi-scientific parlance nowadays. I am using them as the man in the street uses them in everyday life. And the meaning in which we are to take them here will best be made clear by getting directly into touch with the facts to which that meaning is applicable.

Let us begin with the discussion of one of the most elementary concepts in chemistry—atomic weights. The idea of the atom itself dates from the time of the Grecian philosophers. And indeed the word itself, in Greek, means *that which cannot be divided*. The art of measuring atomic weights, however, dates from the discovery of a fundamental principle in stoechiometry. Stoechiometry, by the way, is another Greek word. It is the name which is given to the science of estimating chemical elements. Now, the stoechiometrical principle to which I have just referred is that all chemical compounds result from definite ratios between the weight of one element and another in the compound. For instance, one gram of hydrogen unites with eight grams of oxygen to form water. And if one gram of hydrogen be united with $35 \cdot 5$ grams of chlorine the resulting compound will be hydrochloric acid. If we take one gram of hydrogen as the unit of measurement, we say that eight grams is the *equivalent weight* of oxygen and $35 \cdot 5$ grams the equivalent weight of chlorine. And so for every chemical element in every compound which it can form with another element we can ascertain its equivalent weight. Of course the measurement is based on the choice of hydrogen as a unit, and in that sense of measurement it is somewhat arbitrary. That is not all however. Its validity is restricted to those special elements with which hydrogen combines in order to form a compound. The equivalent weight of oxygen as 8 is valid only in its relation to water. If instead of water we take hydrogen peroxide then the equivalent weight of the oxygen will be 16. In principle there are no grounds whatsoever for preferring one of these numbers to the other. Every element therefore, generally speaking, has a varying equivalent weight. In principle it has as many equivalent weights as there are combinations into which it can enter. If there be an element which does not enter into any known combination then there is no term of reference whereby its equivalent weight can be established. Now the interesting fact is that in the different combinations into which an element may enter with other elements to form a compound, the elements will always be in relation to one another according to their equivalent weight numbers, or a simple multiple of these. This is called the law of multiple proportions, and it states that whenever two elements combine in more proportions than one, the quantities of A, let us say, which combine with a definite quantity of B are connected by a simple multiple. Thus a quantity of chlorine having the equivalent weight of 35·5 combines not only with one gram of hydrogen, to form hydrochloric acid, but also with eight grams of oxygen, to form chloroxide, while it combines with one gram of hydrogen to form hydrochloric acid. Therefore there are *key numbers* which can always be used to describe the proportions of various elements present in the various compounds. To put the matter in a plainer way, in every compound substance the proportional weight of each element may be represented by a fixed number, or by this number multiplied by two, three, four or five and so on. Unless we are to attribute to some inconceivable law of chance this extraordinarily simple and regular scheme into which the various compound substances fit perfectly, we must admit that the idea of equivalent weight must be considered as having an independent significance, irrespective of the combination which the element can make with other elements. Therefore in a certain sense this equivalent weight must be looked upon as something Absolute.

This is what happens in the actual world of fact.

But a difficulty which remained for a long time insoluble in chemistry arose from the fact that some elements are not constant in their valency but may combine with other elements in different ratios, such as hydrogen with oxygen, so that one might take either 8 or 16 as indicating the equivalent weight of oxygen. This difficulty could not be overcome until a new idea was introduced which was foreign to stoechiometry. This idea is contained in Avogadro's Law, which was founded on facts discovered by Gay-Lussac, namely, that two elements in a gaseous state combine with one another not only in definite weight ratios but also in definite volume ratios under equal pressure and temperature. Avogadro's Law states that equal volumes of different gases at the same temperature and pressure contain the same number of molecules, that is to say, the volume of a gram molecule is constant for all gases. Therefore, from the many equivalent weights which might be assigned to each element it was possible to select one definite weight, which was called the molecular weight; because the molecular weight of two gases was found to be in constant ratio to their densities. Here there was no longer any question of chemical reaction but only of chemical substances. Therefore the rule could be applied to elements such as perfect gas, which it is difficult or impossible to combine with other substances.

According to the Avogadrian Law the molecules of chemical elements often enter into the molecules of the combination not with their whole weight but only with a fraction of it. For instance, the molecule of steam is made up of one whole molecule of hydrogen and half a molecule of oxygen, whereas the molecule of hydrochloric acid is made up of half a molecule of chlorine and half a molecule of hydrogen. Therefore from the molecular weight we come to the atomic weight of an element as the smallest fraction which is found in a combination of elements. This atomic weight expresses the relative weights of each species of matter.

Although in Avogadro's Law the concept of atomic weight has a certain absolute significance, at the same time it has quite a relative connotation. The Avogadrian atomic weight is only a relative number. Therefore it cannot be determined except by an arbitrary reference to the atomic weight of some special element or other, such as Hydrogen $= 1$ or Oxygen $= 16$. Without reference to some such given term, the number describing the atomic weight would have no meaning. Therefore it has for a long time been the aim of chemical researchers to free the concept of atomic weight from this restriction and try to give it a wider and more absolute meaning. This problem, however, is not very.important for the practical chemist; because in the chemical analysis of substances there is always the question of relative proportions among the combining elements.

In every science it occasionally happens that there arises a conflict between two classes of people whom I may designate respectively as purists and pragmatists. The former strive always after a perfect co-ordination of the accepted axioms of their science, submitting them to an ever more and more rigid analysis, for the purpose of eliminating every contingent and foreign element. On the other hand, the pragmatists try to amplify the accepted first principles by the introduction of new ideas and thus send out feelers in all directions for the purpose of making progress. They do not mind if the mongrel be mated with the pure-bred, provided something can be achieved through the combination, which overwise could not be achieved. In the science of chemistry also there are purists who set themselves against any attempt to make the concept of atomic weight something more than that of a merely relative number. But there are also leading chemists who find it at least practical to treat the atomic idea as it is treated in mechanical physics, that is to say, to consider the atoms as minute and independent particles occupying definite and measurable dimensions in the molecule, and being either divided or regrouped according as the molecule undergoes chemical changes. During my time in Munich, in the beginning of the eighties, I remember being very much impressed by the polemic that then raged in the university laboratory. Among the puritan chemists the leader was then Hermann Kolbe of Leipzig, who hurled his sacred anathema against the mechanical-atomic interpretation which was involved in the building up of chemical formulas for the constitution of various substances. When results were somewhat slow in being obtained by that process he generally grew all the more violent against the principle adopted. In the circumstances von Baeyer did the wisest thing that could be done. He kept silent and awaited results, until finally success crowned his efforts.

A similar condition of affairs was reproduced recently when the controversy arose over the atom model suggested by Niels Bohr, which indeed demands a far greater concession on the part of orthodox theorists than the earlier hypothesis of the atomic structure of chemical elements.

On the philosophical side also there are purists who have maintained a long-standing attitude of opposition to the atomic theory. Ernst Mach was the most outstanding leader of this school. During his life he never seemed to tire of using the weapon of conceptual analysis, and occasionally also his irony, for the purpose of discrediting the rather naïve and rudimentary views of those who then championed the atomic principle. He believed that the revival of the old atomic doctrine and the dressing of it in modern form signified a retrogression, and hindered rather than helped the philosophical development of modern physics.

Ludwig Boltzmann, as leading representative of the atomic physicists, boldly endeavoured to hold his ground against Mach; but the contest was rather difficult from his side, because the purist sticks to his logical weapons. He takes his stand on logical deductions from the accepted principles of science, whereas the pragmatist scientist is striking out into new ground; and in order to open that up he must break away from the logical line of the old ideas. The pragmatist must face failure again and again, and is always open to the jibes of the orthodox "I told you so." What the puritan objects to is the introduction of new ideas and theorems from outer sources, especially while those are in the stage of not having produced any results in practice. Now, no theorem or working hypothesis can arise ready-made, like Pallas Athene from the head of Zeus. Every hypothesis which eventually has proved to be useful and to have led to valuable discoveries at first occurred only vaguely to the mind of its inventor. When Archimedes jumped out of his bath one morning and cried *Eureka* he obviously had not worked out the whole principle on which the specific gravity of various bodies could be determined; and undoubtedly there were people who laughed at his first attempts. That is perhaps why most scientific pioneers are so slow to disclose the nature of their first insights when they believe themselves to be on the track of a new discovery. They would have to stand against the massed batteries of the purists, which would not be a very advisable position for anyone to take up who has to follow the lead of his own instinct painfully and painstakingly and refuse to be discouraged when his attempts end in failure. For every hypothesis in physical science has to go through a period of difficult gestation and parturition before it can be brought out into the light of day and handed to others, ready-made in scientific form so that it will be, as it were, fool-proof in the hands of outsiders who wish to apply it.

Even when a scientific theory has established its

M

right to existence by reason of the results it has produced, the purist often takes a long time to come round. And that is because the success of a new theory in physics cannot be decided according to its logical consistency with accepted notions, but rather by the test whether or not it explains and co-ordinates certain facts already ascertained, but which cannot be explained on any other grounds except that of the new hypothesis. Of course the purists have always the old refuge to fall back upon. They appeal to the element of chance. And on that stand some of them will remain, while others will take up an intermediate position of qualified scepticism; but the pragmatist finds that the hypothesis in question has worked out a clear solution of certain puzzles and he accepts it for what it does. Instead of looking backwards he begins to look forward with a view to finding whether the hypothesis may not be applied in other directions also. It was thus with the fate of the quantum hypothesis, for instance. It was originally formulated to explain a puzzle of radiation which had long existed; but in the hands of Einstein it was soon applied to explain the constitution of light and, in the hands of Niels Bohr, to explain the structure of the atom.

It was just in this way that the existence of an absolute atomic weight came to be finally established. Here I need not go into details to tell how so many lines of research led finally to the discovery of the absolute atomic weight. Among these many lines I may mention the development of the kinetic theory for gases and fluids, the laws governing the radiation of heat and light, the discovery of the cathode rays and radioactivity, and the measurement of the elementary electrical quantum. To-day no physicist would question the fact that the weight of an atom of hydrogen, setting aside the unavoidable errors of measurement, amounts to 1·649 quadrillionths of a gram. The value of this number is entirely independent of the atomic weight of other chemical elements, and in this sense it can be called an absolute quantity.

All this of course is already a matter of common knowledge. And I have mentioned it here in order to illustrate a characteristic feature in the development of scientific research. This phenomenon shows itself under the most varied circumstances. Axioms are instruments which are used in every department of science, and in every department there are purists who are inclined to oppose with all their might any expansion of the accepted axioms beyond the boundary of their logical application.

I shall now suggest another case for consideration. But this is by no means so simple as that which I have already treated. In fact it is still the centre of contention.

Let us begin with a concept of energy. The term "energy" represents the work that can be done by forces acting on matter. And the Principle of the Conservation of Energy, which was formulated in the middle of last century, was a development from the concept of force in Newtonian mechanics. According to the Principle of the Conservation of Energy, in every mechanical process the amount of energy which the moving force puts into the body moved is compensated for by a loss of potential energy on the part of the acting force. Two kinds of energy were thus recognized, namely potential energy and kinetic energy, the former being the energy possessed by bodies at rest and the latter being the energy of moving bodies. There is no such thing then as absolutely lost energy, but only a change from one kind of energy to another. And the loss sustained by one kind of energy, the potential, is compensated for by the gain in the other kind of energy, the kinetic. In this connection the purist might reasonably maintain that the formulation of the Principle of the Conservation of Energy is valid only for a difference of energy, and that the concept of energy does not refer to the state of a body or, as we say in scientific language, the state of a physical system, but rather to a change in that state. Therefore the energy value remains an indefinite superadded factor. And the question of its measurement would have no meaning in physical science. It would have the same relation to the physicist as the altitude above sea-level would have for the architect who is building a house. It is not the latter's business to bother himself about this altitude. He has to confine himself to the altitude of the house itself and that of the various floors of which it is composed. Such is the objection that a purist might urge.

His standpoint would be quite sound if the Principle of the Conservation of Energy were the only axiom employed in physical science. But this is not the case. And therefore we cannot reject off-hand the suggestion that it may be well to introduce into the concept of energy that of another axiom, if the result would be that the state of a physical picture here and now could thus be fully determined. If we could do that then it is obvious that the concept of energy would be very much simplified by the addition of something else to the Principle of Conservation. As a matter of fact that is what has been done to-day. For any physical system in a given state we can find a definite expression for the magnitude of its energy, without any superadded factor whatsoever.

Let us first take electromagnetic energy in a vacuum. Here there is an axiom which establishes the absolute value of that energy. It states that the energy of an electromagnetic neutral field is equal to Zero. This law is neither obvious in itself nor can it be deduced from the Principle of the Conservation of Energy. Only a few years ago Nernst formulated the hypothesis that in the so-called neutral field there is a certain stationary energy radiation of tremendous magnitude. This is called the radiation of the Zero point. It cannot be detected in the observation of ordinary processes because it streams through all bodies equally, just as the pressure of the atmosphere represents a very important force which plays no part in most of the movements that we observe, because the pressure is equal in all directions. Such a radiation hypothesis is perfectly reasonable, and its validity can be decided upon only by the question of what results follow from its application. For this application, however, it is absolutely necessary to furnish a special reference

system that is immobile, namely that in which the
Zero radiation is equal in all directions. Through the
absolute energy of the neutral field the absolute
energy of every other electromagnetic field is thereby
established.

Coming now to the energy of matter, for this we
can also obtain a definite absolute value. But the
energy of a body at rest is not equal to Zero as might
probably be imagined, following the analogy of the
electromagnetic neutral field. The energy of a body
at rest is equal to its mass multiplied by the square
of the velocity of light. This is the so-called rest energy
of the body, and is caused by its mechanical constitution and its temperature. If the body be set in motion
by some force this energy value, which is of an enormous amount, does not make itself felt because the
phenomenon of motion here arises from only a differentiation of energy. Such a conception could never have
arisen from the energy principle itself. As a matter of
fact it arises from the special theory of relativity, and
it is a remarkable coincidence that it is just the theory
of relativity which has led to the determination of an
absolute value for the energy of a physical system.
This apparent paradox is explained by the simple
fact that in the relativity theory there is the question
of dependence on the reference system selected,
whereas here there is the question of dependence on
the physical state of the body under observation.

"Doesn't it in reality sound quite nonsensical to
say that the energy of an atom of oxygen is sixteen
times greater than that of an atom of hydrogen?" the

purist might ask. We might answer that there would be no sense in such a statement if we could not speak of the hypothetical transformation of oxygen into hydrogen without involving a logical contradiction in the thought itself. But the idea of oxygen being one day changed into hydrogen does not involve any logical contradiction. Now, it is a mistake in these matters to speak of something as nonsensical unless it can be shown to be logically incoherent; and it would therefore seem more advisable to wait and see whether a day may not come when the problem of this transformation of oxygen into hydrogen may assume a reasonable significance. There are already signs that this time is at hand.

As in the case of electromagnetic and kinetic energy so too in all departments of physics, mechanics as well as electrodynamics, the movement has been away from dealing with differentials of energy towards dealing with absolute values of it. And this direction has invariably led to important results. When considering the phenomenon of heat radiation, for instance, it was always the strict rule to deal only with the difference between the radiation absorbed and that emitted; because all the heat rays that a body absorbs it can also give out. But in the theory of Prevost these two processes were separated from one another and each of them given an independent meaning. In galvanism only the potential difference was measured; but the absolute value of the potential was also recognized because the potential energy of all electric charges at infinite distances was declared to be equal to Zero. For the emission of monochromatic radiation in the case of an atom the measurement of the frequency emitted gave only a difference of the atomic energy before and after the emission. But by first separating the two factors of this difference—the so-called terms—and then examining each separately, Niels Bohr and Arnold Sommerfeld were able to discover a clue for the solution of the mystery, Niels Bohr in the case of visible rays and Arnold Sommerfeld for the Roentgen rays.

It is not, however, merely in its dealings with the problem of energy that progress from the differential to the integral is characteristic of physical science. We find the same feature showing itself in every other branch of physical research. Thus the older elasticity theory of body force is now referred back to surface forces. In electrodynamics electric and magnetic ponderometer forces are resolved into the so-called Maxwellian tension. The thermodynamic measurements of temperature and pressure are resolved into the thermodynamical potential. In each of these cases the progress signifies a new stage in the evolution of theoretical physics.

But there is one evolutionary struggle going on which deserves a little more detailed notice because it is still in an undecided phase. It is the problem of trying to find an absolute value for entropy. In the original definition of entropy put forward by Rudolf Clausius, if we are to measure the entropy of a body there must be a reversible process of some kind to enable us to determine the difference of entropy between the initial state and the final state of the process. In the light of this theory the concept of entropy originally referred not to a state but rather to a change of state, exactly as was the case in regard to atomic weight and energy. Indeed the earlier scientific notion was that the concept of entropy had a physical significance only where there could be a reversible process. It did not take long, however, before a broader concept was put forward and entropy began to be looked upon as a characteristic or inherent quality in the state of a body here and now. In this new way of looking at the case, however, there still remained an undefined additive constant, because one could still measure only the difference of entropy. Were we to follow the lead suggested by the Einstein experiments, and base the concept of entropy on the statistical laws governing the oscillations of a physical picture in relation to its thermodynamic state of equilibrium, even then we should only arrive at a measurement of differences, involved in a change of entropy, but never at the absolute value of entropy itself.

Is there then any way whereby we can hope to find an absolute value for entropy as has been found for energy? I do not think that the question can be answered on the basis of an analogy between these two cases. When such suggestions come to the fore I am always inclined to take my stand with the purists, who hold that it is senseless to try to arrive at the values of both *termini* from the value of the difference. If we are to keep our outlook clear we must always be very careful as to what can or cannot be deduced from a definition. In this regard the criterion of the purists is indispensable. We must do them the honour of saying that they are the conscientious wardens of order and purity in scientific methods. There is nothing more seductively dangerous in scientific work than the introduction of extraneous analogies into the problem at issue. That is a warning which needs to be sounded to-day even more insistently than before. But at the same time we must bear in mind the fact that physics is not a deductive science, and that its body of first principles is by no means fixed and unalterable. If a new axiom be suggested which we might introduce, then instead of rejecting it at once it ought to be put into quarantine, as one might say, and examined on its own merits for a clean bill of health. That clean bill of health which will give it a right to citizenship in physical science must be drawn up entirely free from prejudice as to the alien status of the axiom. The claim of the axiom must be adjudicated on the grounds of its ability to serve the cause of science in some direction where service is needed, and where the native axioms are unable to discharge such service. Once the new axiom has shown that it can solve hitherto insoluble problems, or at least produce a working hypothesis for their explanations, then it has a perfect right to be admitted.

Before indicating a definite line along which the question I have given above may eventually be answered, I will call attention to the difference between reversible and irreversible processes, and from this we shall understand the Boltzmann hypothesis which would suggest the answer. Suppose we take a piece of iron heated to a very high temperature and plunge it into a vessel of cold water. The heat of the iron will pass to the water until both iron and water are of an equal temperature. This is called thermal equilibrium, which results after all such cases of disturbance if there be nothing to prevent the conduction of the heat.

Now let us take two vertical tubes of glass which are open at the upper ends and have the lower ends connected by a piece of rubber tubing. If we pour some heavy liquid such as mercury into one of the glass tubes the liquid will flow through the rubber into the second tube and rise in it until the level of the surfaces in both tubes is the same. Now supposing we lift one of the tubes somewhat, the level is disturbed; but the fluid will fall back immediately when we replace the tube and will again be the same height in both. Between this instance and that of the bar of iron in the vessel of water there is a certain analogy. In each case a certain difference brings about a change. In the case of the tube which we raise a little higher than the other there is a change of level, and in the case of the iron and water there is at the moment of immersion a difference between the temperatures. If in each case we allow the total mass to rest sufficiently long the differences will disappear and a condition of equilibrium will result.

As a matter of fact the analogy between these two cases is only apparent. All experiments which have been made warrant us in definitely asserting that the action of the liquid in the tubes follows a dynamical law, but that the energy of temperature follows a statistical law.

To understand this apparent paradox we must remember that the sinking of the heavy liquid is a necessary consequence of the Principle of Conservation of Energy. For if the liquid at a higher level were to rise still higher irrespective of any external agency, and the liquid of the lower level to sink still lower, energy would be created out of nothing. That is to say, new energy would appear and thus be entirely contrary to the principle. The temperature case is different. Heat could flow in the reverse process from cold water to hot iron, and the Principle of Conservation of Energy still hold good; because heat itself is a form of energy, and the principle only demands that the quantity of heat given up by the water be equal to that absorbed by the iron.

Now the two operations show the following different characteristics. The falling liquid moves faster the further it falls. When the level in one tube corresponds to the level in the other the liquid does not come to rest, but moves beyond the equilibrium point on account of its inertia, so that the liquid originally at the higher level is now at a lower level than that rising in the corresponding tube. The velocity of the falling liquid will gradually sink to zero in tube No. 1 and then the reverse process sets in, that is to say the lowering of the level in tube No. 2. If loss of kinetic energy at the air surface, and that due to friction at the walls of the tube could be eliminated, the liquid would oscillate upwards and downwards indefinitely over and under its position of equilibrium. Such a process is called reversible.

Now in the case of heat the condition is quite otherwise. The smaller the difference of temperature between the hot iron and the water the slower is the transmission of heat from the one to the other, and calculation shows that an infinitely long time passes before an equal temperature is reached. This means that there is always some difference of temperature no matter how much time be allowed to elapse. There is no oscillation of heat therefore between two bodies. The flow is always in one direction and therefore represents an irreversible process.

This difference between reversible and irreversible processes is fundamental in physical science. Reversible processes include gravitation, mechanical and electrical oscillations, sound waves and electromagnetic waves. Irreversible processes are found in the conduction of heat and electricity, radiation and all chemical reactions in so far as the velocity is ascertainable. It was to explain this case that Clausius formulated his second law of thermodynamics. The significance of the law is that it ascribes direction to each irreversible process. It was L. Boltzmann however who introduced the atomic theory here and thus explained the meaning of the second law, and at the same time of all irreversible processes which hitherto had presented difficulties that could not be explained in classical dynamics.

According to this atomic theory the thermal energy of a body is the sum-total of a small, rapid, and unregulated movement of its molecules. The temperature corresponds to the medium kinetic energy of the molecules, and the transfer of heat from a hotter to a colder body depends upon the fact that the kinetic energies of the molecules are averaged because of their frequent collision with one another. It must not be supposed however that when two individual molecules strike together the one with the greater kinetic energy is slowed down and the other accelerated, for if—to take an example—a rapidly moving molecule of one system is struck obliquely by a slower moving molecule its velocity is increased while that of the slower moving molecule is still further diminished. But, taken on the whole, unless the circumstances are quite exceptional the kinetic energies must mix to a certain amount, and this mixing is what appears as an equalizing of the temperature of the two bodies.

Boltzmann however did not press his hypothesis very strongly before the notice of scientists and there was great hesitancy about accepting it, but nowadays it is fully accepted. It is now generally agreed that heat movement of molecules and conduction of heat, like all other irreversible phenomena, do not obey dynamical laws but statistical laws. The latter are the laws of probability.

Now in the case under consideration it is not at all difficult to say what the idea is that lies behind the assumption of an absolute value for entropy. And if a new axiom can serve that idea we ought to admit it. As to the idea of absolute value for entropy, if we follow Boltzmann and consider entropy as a measure for thermodynamic probability, then when a physical state such as a volume of gas, with various degrees of freedom and endowed with a definite energy, has reached a condition of thermodynamic equilibrium, the entropy in such a case will be nothing more than the number of the multiform states which such a system can assume under given conditions. And if the entropy thus considered possesses an absolute value this means that the number of possible states under the given conditions is quite definite and finite.

At the time of Clausius and Helmholtz and Boltzmann such an assertion would have been considered entirely out of the question. The differential equations of classical dynamics were then looked upon as the sole fundamentals of physical science. Therefore it was necessary to consider physical states as continuous, and all possibilities of change as infinite in their measurable quantities. Since the introduction of the quantum hypothesis the state of affairs is different, and I feel that we have not long to wait before it will be possible to speak in quite a different way of a definite number of possible states and of absolute measures of entropy corresponding to them, without thereby running up too violently against the accepted physical notions of the time. Indeed the new quantum axiom has already produced results that can favourably compare with the most fruitful theories of the past. In the case of radiant heat it has led to the formulation of laws of energy which explain the normal spectrum. In the laws of thermodynamics it has found its expression in the theory established by W. Nernst, which has been corroborated on many sides; and the basis of the quantum hypothesis has been so far expanded that from it we can deduce not only the existence but also the numerical values of the so-called chemical constants. In regard to the constitution of the atom the ideas of Niels Bohr have been the starting-point for the establishment of the so-called stationary electronic orbits, and thus the ground was prepared for solving the riddle of the spectroscopic phenomena. Indeed, unless all signs be misleading a process seems to be developing which may be called the reduction of all physical theories to arithmetical terms, because a large number of physical dimensions which hitherto had been looked upon as continuous have been shown under the microscopic examination of a sharper analysis to be discontinuous and numerable. Along these lines the measurements which have been arrived at by L. S. Ornstein, the head of the Physical Institute at Utrecht, are indicative. These measurements show that the ratio of intensity of the components of spectral multiplets can be given in simple integral numbers. And Max Born's interesting attempt to supplant the differential calculus of physical mechanics by equations of finite differences points in the same direction.

The outstanding cases that I have here chosen point to a definite *Drang* or fundamental urge which seems to characterize the advance of physical science.

In these cases the movement has undoubtedly been from the relative to the absolute. Now comes the question: How far can we say that this advance is definitely characteristic of the progress of physical science as a whole? It would be saying too much, perhaps, if I were to answer the question by an unqualified affirmative. Indeed I can easily imagine that some of my readers may be of the opposite view, and may already be thinking in their own minds that this chapter could be written in the reverse direction and called "From the Absolute to the Relative." They certainly would find material at hand which at least on the surface offers tempting ground to stand upon. It might, for instance, be urged that the concept of atomic weight could be taken as pointing in a direction contrary to that which I have suggested. My imaginary opponent might say that the numeral which I have indicated as representing the absolute weight of an atom is by no means absolute. In view of the fact that an element generally possesses several isotopes with a different atomic weight, the measured atomic weight presents a more or less contingent addition which is a sort of average value, that is quite dependent on the ratio of the various isotopes in the compound under analysis. Even if we were to take only one single isotope into consideration, from the standpoint of our present knowledge it would be quite unscientific to consider this as something absolute. The most modern opinion, which is backed up by the Rutherford experiment of bombarding the nucleus of the atom, would seem to be in the direction of reviving Prout's hypothesis and referring the constitution of all chemical elements to the basic atom of hydrogen. Therewith the concept of atomic weight would fundamentally be a relative number. Having thus gained what at least appears to be a signal victory in this one instance, my opponent might play his trump card and throw the Einstein General Theory of Relativity on the table. He might very well urge that to talk of the concepts of space and time as something absolute belongs to the past and signifies retrogression rather than progress. In other words, one of the most signal advances in modern physics is stamped with the idea of the relative rather than the absolute.

The first and most obvious reply to such a criticism is to call attention to the danger of applying scientific terms to facts and meanings for which they were never intended. I have already shown how the theory of relativity has actually led to the discovery of an absolute measure by which the energy of a body at rest may be formulated. Therefore it is clear that the term, *Relativity*, does not refer to physics as a whole and must not be taken out of its special scientific context. It would be quite superficial to take the relativity of time and space, and halt firmly within the confines of that concept without asking whither it leads. As a matter of fact the concept of relativity is based on a more fundamental absolute than the erroneously assumed absolute which it has supplanted. Over and over again in the history of science it has happened that concepts which at one time were looked upon as absolute were subsequently shown to be only of relative value; and this is exactly what has happened in regard to the former concept of space and time. But when an absolute concept is thus relativized, this does not mean that the quest of the absolute becomes eliminated from scientific progress. It rather means that a more fundamental concept takes its place and a more fundamental advance is thus achieved. If we admit the concept of relativity at all we must admit the acceptance of an absolute, because it is out of this that the relative concept as such arises. Supposing, for instance, a scientific researcher worked for years and years on the problem of discovering the cause of some special event in nature and found all his efforts baffled, would he thereby be justified in declaring that the event has no cause at all? The fact is that we cannot relativize everything any more than we can define and explain everything. There are fundamentals that cannot be defined or explained, because they form the bedrock of all our knowledge. Every definition must necessarily rest on some concept which does not call for definition at all. And it is the same with every form of proof. We cannot define a thing except in terms that are already known and accepted, and we cannot prove anything except from something that is already admitted. If we wish to establish a truth by the inductive method it must be on the basis of accepted facts. And if we wish to establish a truth by the process of deductive reasoning the principle from which the deduction proceeds must be accepted as absolute. Therefore the relativist concept must necessarily have the concept of the absolute as its foundation. If we once remove the absolute, then the whole relativist theory will fall to the ground, just as an overcoat would fall if the peg on which it hangs should disappear. These considerations are quite sufficient, I think, to suggest the reply which might be given to the counter argument of my imaginary disputant.

If eventually it should turn out possible to refer the atomic weights of all elements to the atomic weight of hydrogen, then we should have achieved one of the most fundamental results in the history of the scientific investigation of matter. The significance of it would be that in the light of this explanation matter could be proved to have one simple origin. Then the two factors of the hydrogen atom, namely, the positively charged hydrogen nucleus (the so-called proton) and the negatively charged electron, together with the elemental quantum of action, would represent the foundation-stones on which the structure of the physical world is built. Now these quantities should be considered as absolute as long as they do not depend upon one another or something outside of them. There we should have the absolute once again, only at a higher level and in a simpler form. If we like to unroll the thread of this thought a little further, we might ask what is the foundation on which the great relativist theory is built? Einstein explained that our concepts of space and time, which were recognized by Newton and Kant as absolute forms of all knowledge, really possessed only a relative significance, inasmuch as they depended on an arbitrary selection of the reference system and the means of measurement. It is a familiar fact that we cannot observe the motion of any body without reference to some other body. It was to meet this difficulty that Newton adopted the hypothesis of absolute space. The "fixed" stars were used to define absolute space. The stars, however, are not fixed even relatively to one another. Therefore the concept of absolute space and the reference points according to which it was "fixed" were quite arbitrary. This explanation goes perhaps to the deepest root of our scientific thought. If from space and time we should take away the concept of the absolute, this does not mean that the absolute is thereby banished out of existence, but rather that it is referred back to something more fundamental. As a matter of fact, this more fundamental thing is the four-dimensional manifold which is constituted by the welding together of time and space into a single continuum. Here the standard of reference and measurement is independent of arbitrary choice and is absolute.

It only takes a little reflection to realize the fact that the much misunderstood relativity theory by no means gets rid of the absolute but, on the contrary, that it has brought out the absolute into sharper definition, inasmuch as it points out how, and how far, physical science is based on the existence of an absolute in the outer world. If we should say, as several epistemologists do, that the absolute is to be found only in the individual's sensory data of perception, then there ought to be as many kinds of physical science as there are physicists, and we should be utterly unable to explain how it is that up to now each discoverer in physical science has been standing on the shoulders of his predecessors, as it were, and has taken their findings as the basis of his work. Indeed it is exclusively on the basis of cooperative labour and the acceptance by others of the findings of the various individual researchers, that we can explain the structure of physical science as we have it to-day. That we do not construct the external world to suit our own ends in the pursuit of science, but that *vice versa* the external world forces itself upon our recognition with its own elemental power, is a point which ought to be categorically asserted again and again in these positivistic times. From the fact that in studying the happenings of nature we strive to eliminate the contingent and accidental and to come finally to what is essential and necessary, it is clear that we always look for the basic thing behind the dependent thing, for what is absolute behind what is relative, for the reality behind the appearance and for what abides behind what is transitory. In my opinion, this is characteristic not only of physical science but of all science. Further, it is not merely a characteristic of all kinds of human endeavour to attain to the knowledge of any subject, but it is also characteristic of those branches of human effort that strive to formulate ideas of the good and the beautiful.

Here I am going wide of my purpose; for the plan I had in mind at the beginning of this essay was not to make assertions and then prove them, but rather to call attention to certain actual changes which have taken place in the course of scientific development and allow the bare presentation of facts to leave its own impression on the mind of the reader.

Before closing I should like to raise the most difficult question of all. It is this: How can we say that a scientific concept, to which we now ascribe an absolute character, may not at some future date show itself to have only a certain relative significance and to point to a further absolute? To that question only one answer can be given. After all I have said, and in view of the experiences through which scientific progress has passed, we must admit that in no case can we rest assured that what is absolute in science to-day will remain absolute for all time. Not only that, but we must admit as certain the truth that the absolute can never finally be grasped by the researcher. The absolute represents an ideal goal which is always ahead of us and which we can never reach. This may be a depressing thought; but we must bear with it. We are in a position similar to that of a mountaineer who is wandering over uncharted spaces, and never knows whether behind the peak which he sees in front of him and which he tries to scale there may not be another peak still beyond and higher up. Yet it is the same with us as it is with him. The value of the journey is not in the journey's end but in the journey itself. That is to say, in the striving to reach the goal that we are always yearning for, and drawing courage from the fact that we are always coming nearer to it. To bring the approach closer and closer to truth is the aim and effort of all science.

Here we can apply the saying of Gotthold Ephraim Lessing. "Not the possession of truth but the effort in struggling to attain to it brings joy to the researcher." We cannot rest and sit down lest we rust and decay. Health is maintained only through work. And as it is with all life so it is with science. We are always struggling from the relative to the absolute.

EPILOGUE

A SOCRATIC DIALOGUE

INTERLOCUTORS: EINSTEIN—PLANCK—MURPHY

Note:—The following is an abridgment of stenographic reports made by an attendant secretary during various conversations

MURPHY: I have been collaborating with our friend, Planck, on a book which deals principally with the problem of causation and the freedom of the human will.

EINSTEIN: Honestly I cannot understand what people mean when they talk about the freedom of the human will. I have a feeling, for instance, that I will something or other; but what relation this has with freedom I cannot understand at all. I feel that I will to light my pipe and I do it; but how can I connect this up with the idea of freedom? What is behind the act of *willing* to light the pipe? Another act of willing? Schopenhauer once said: *Der Mensch kann was er will; er kann aber nicht wollen was er will* (Man can do what he wills but he cannot will what he wills).

MURPHY: But it is now the fashion in physical science to attribute something like free will even to the routine processes of inorganic nature.

EINSTEIN: That nonsense is not merely nonsense. It is objectionable nonsense.

MURPHY: Well, of course, the scientists give it the name of indeterminism.

EINSTEIN: Look here. Indeterminism is quite an illogical concept. What do they mean by indeterminism? Now if I say that the average life-span of a radioactive atom is such and such, that is a statement which expresses a certain order, *Gesetzlichkeit*. But this idea does not of itself involve the idea of causation. We call it the law of averages; but not every such law need have a causal significance. At the same time if I say that the average life-span of such an atom is indetermined in the sense of being not caused, then I am talking nonsense. I can say that I shall meet you to-morrow at some indetermined time. But this does not mean that time is not determined. Whether I come or not the time will come. Here there is question of confounding the subjective with the objective world. The indeterminism which belongs to quantum physics is a subjective indeterminism. It must be related to something, else indeterminism has no meaning, and here it is related to our own inability to follow the course of individual atoms and forecast their activities. To say that the arrival of a train in Berlin is indetermined is to talk nonsense unless you say in regard to what it is indetermined. If it arrives at all it is determined by something. And the same is true of the course of atoms.

MURPHY: In what sense then do you apply determinism to nature? In the sense that every event in nature proceeds from another event which we call the cause?

EINSTEIN: I should hardly put it that way. In the first place, I think that much of the misunderstanding encountered in all this question of causation is due to the rather rudimentary formulation of the causal principle which has been in vogue until now. When Aristotle and the scholastics defined what they meant by a cause, the idea of objective experiment in the scientific sense had not yet arisen. Therefore they were content with defining the metaphysical concept of cause. And the same is true of Kant. Newton himself seems to have realized that this pre-scientific formulation of the causal principle would prove insufficient for modern physics. And Newton was content to describe the regular order in which events happen in nature and to construct his synthesis on the basis of mathematical laws. Now I believe that events in nature are controlled by a much stricter and more closely binding law than we suspect to-day, when we speak of one event being the *cause* of another. Our concept here is confined to one happening within one time-section. It is dissected from the whole process. Our present rough way of applying the causal principle is quite superficial. We are like a child who judges a poem by the rhyme and knows nothing of the rhythmic pattern. Or we are like a juvenile learner at the piano, just relating one note to that which immediately precedes or follows. To an extent this may be very well when one is dealing with very simple and primitive compositions; but it will not do for the interpretation of a Bach Fugue. Quantum physics has presented us with very complex processes and to meet them we must further enlarge and refine our concept of causality.

MURPHY: You'll have a hard job of it, because you'll be going out of fashion. If you will permit me to make a little speech I shall do so, not so much because I like to listen to my own talk, though of course I do—what Irishman doesn't?—but rather because I should like to have your reactions to it.

EINSTEIN: *Gewiss.*

MURPHY: The Greeks made the workings of fate or destiny the basis of their drama; and drama in those days was a liturgical expression of the profound irrationally perceiving consciousness. It was not merely a discussion, like a Shavian play. You remember the tragedy of *Atreus,* where fate, or the ineluctable sequence of cause and effect, is the sole simple thread on which the drama hangs.

EINSTEIN. Fate, or destiny, and the principle of causation are not the same thing.

MURPHY. I know that. But scientists live in the world just like other people. Some of them go to political meetings and the theatre and mostly all that I know, at least here in Germany, are readers of current literature. They cannot escape the influence of the *milieu* in which they live. And that *milieu* at the present time is characterized largely by a struggle to get rid of the causal chain in which the world has entangled itself.

EINSTEIN. But isn't mankind always struggling to get rid of that causal chain?

MURPHY. Yes, but that is not to the point just at the moment. Anyhow I doubt if the politician ever contemplates the consequences of the causal sequence he sets afoot by his foolishness. He is too nimble himself and can slip out through the links. Macbeth was not a politician. And that is where he failed. He realized that the assassination could not trammel up the consequence. But he did not think of how to escape from the sequential shackles until it was too late. And this is because he was not a politician. My point here is that there is a universal recognition at the moment of this inexorable sequence. People are realizing what Bernard Shaw told them long ago —which of course had been told on innumerable occasions previously—when he wrote *Caesar and Cleopatra*. You remember Caesar's speech to the Queen of Egypt after her orders to slay Photinus had been carried out, though Caesar had guaranteed his safety.

"Do you hear?" says Caesar. "Those knockers at your gate are also believers in vengeance and in stabbing. You have slain their leader; it is right that they shall slay you. If you doubt it, ask your four councillors here. And then in the name of right shall I not slay them for murdering their Queen, and be slain in my turn by their

countrymen as the invader of their Fatherland?
Can Rome do less than slay these slayers too, to
show the world how Rome avenges her sons and
her honour? And so, to the end of history, murder
shall breed murder, always in the name of right
and honour and peace, until the gods are tired of
blood and create a race that shall understand."

People realize this terrible truth nowadays, not
indeed because they see that blood will have
blood but because they see that in robbing your
neighbour you rob yourself; for robbery will have
robbery just as blood will have blood. The so-
called victors in the world war robbed the van-
quished and they now know that in doing so
they robbed themselves. So now we have a
condition of all-round misery. People at large see
that; but they haven't the courage to face it and
they race, like Macbeth, to the witches' cauldron.
In this case unfortunately science is one of the
ingredients thrown into the cauldron to give
them the solvent they are looking for. Instead of
boldly admitting the mess, the tragedy, the
crime, everybody wants to try to prove himself
innocent, and looks for the proof by trying to find
an alibi for the consequences of his own deeds.
Look at that string of hungry people coming to
your door every day for bread. Able-bodied
fellows who want to exercise man's privilege,
which is to work. You have them also parading
the streets of London, with their Distinguished
Conduct Medals on their breasts, shouting for

bread. And you have the same in New York and Chicago and Rome and Turin. The comfortable person who sits in his easy chair says to himself, "This has nothing to do with us." And he says that because he knows it has. Then he takes up his popular writers of physics and gives a sigh of contentment when he is told that nature knows no such thing as the law of consequences. What more do you want? Here is Science; and Science is the modern counterpart of religion. It is your comfortable *bourgeois* who has endowed scientific institutions and laboratories. And, say what you will, scientists would not be human if they did not, at least unconsciously, share in the same spirit.

EINSTEIN: *Ach das kann man nicht sagen.*

MURPHY: Yes. That can man very well say. You remember your own picture of the self-seekers in the temple of science, who you admit have built even a great portion of the structure, while you acknowledge that only a few have found favour with the angel of God. I am inclined to think that the struggle of science at the present moment is the effort to keep its thought-scheme clear of the confusion which the popular spirit would bring into it. It is much the same struggle as the old theologians had. At the Rennaissance however they succumbed to the fashion of the time and introduced foreign ideas and methods into their science, which finally resulted in the scholastic break-up.

The decline of scholasticism dates from the time when the mob started running after the philosophers and theologians. Remember how they rushed helter-skelter to hear Abelard in Paris, though it is obvious that they could not understand his distinctions. Public flattery was more the cause of his downfall than any merely private influences. He would not have been human if he had not been tempted to think himself above his science, and he succumbed to the temptation. I am not so sure that many scientists are not in his place to-day. Some of the glistening webs of fancy that they weave seem very much akin to the sophistic distinctions of the scholastic decadence.

The older philosophers and theologians were aware of this danger and they contrived to offset it. They had their esoteric bodies of doctrine which were disclosed only to the initiated. We have the same sort of protection evidenced in other branches of culture to-day. The Catholic Church has wisely maintained its ritual and dogmas within the forms and formulations of a language which the populace does not understand. The sociologists and financial experts have a jargon that is all their own and it saves them from being found out. The majesty of the law is upheld in like manner and the medical craft could not survive if it prescribed its medicines and described its diseases in the vernacular. But all these do not matter because none of all these

sciences or arts or crafts are vital. Physical
science is organically vital at the moment and for
that reason it seems to be suffering from——

EINSTEIN: But I can think of nothing more objectionable than the idea of science for the scientists.
It is almost as bad as art for the artists and
religion for the priests. There is certainly something in what you say. And I believe that the
present fashion of applying the axioms of physical
science to human life is not only entirely a
mistake but has also something reprehensible in
it. I find that the problem of causality which is
to-day under discussion in physics is not a new
phenomenon in the field of science. The method
which is being used in quantum physics has
already had to be applied in biology, because the
biological processes in nature could not in themselves be traced so that their connection would
be clear, and for that reason biological rules have
always been of a statistical character. And I do
not understand why so much pother ought to
be made if the principle of causation should
undergo a restriction in modern physics, for this
is not a new situation at all.

MURPHY: Of course it has not brought about any new
situation; but biological science is not vital in the
way that physical science is vital at the moment.
People are no longer very much interested
whether we were descended from monkeys or
not, except certain animal enthusiasts who think
the idea rather rough on the monkey. And there

o

is not that public interest in biology such as there was in the time of Darwin and Huxley. The centre of gravity of the public interest has shifted to physics. That is why the public reacts in its own way to any new formulation in physics.

EINSTEIN: I am entirely in agreement with our friend Planck in regard to the stand which he has taken on this principle, but you must remember what Planck has said and written. He admits the impossibility of applying the causal principle to the inner processes of atomic physics under the present state of affairs; but he has set himself definitely against the thesis that from this *Unbrauchbarkeit* or inapplicability we are to conclude that the process of causation does not exist in external reality. Planck has really not taken up any definite standpoint here. He has only contradicted the emphatic assertions of some quantum theorists and I agree fully with him. And when you mention people who speak of such a thing as free will in nature it is difficult for me to find a suitable reply. The idea is of course preposterous.

MURPHY: You would agree then, I imagine, that physics gives no ground whatsoever for this extraordinary application of what we may for convenience' sake call Heisenberg's principle of indeterminacy.

EINSTEIN: Of course I agree.

MURPHY: But then you know that certain English physicists of very high standing indeed and at the same time very popular have promulgated

with emphasis what you and Planck call, and
many others with you, unwarranted conclusions.

EINSTEIN: You must distinguish between the physicist
and the *littérateur* when both professions are
combined into one. In England you have a
great English literature and a great discipline
of style.

MURPHY: Literature abhors that *amor intellectualis*
of logical truth which is the scientist's passion.
Perhaps the English scientist changes his colour
in literary pastures so that, like the caterpillar on
the leaf, he may escape detection.

EINSTEIN: What I mean is that there are scientific
writers in England who are illogical and romantic
in their popular books, but in their scientific
work they are acute logical reasoners.

What the scientist aims at is to secure a
logically consistent transcript of nature. Logic
is for him what the laws of proportion and per-
spective are to the painter, and I believe with
Henri Poincaré that science is worth pursuing
because it reveals the beauty of nature. And here
I will say that the scientist finds his reward in
what Henri Poincaré calls the joy of comprehen-
sion, and not in the possibilities of application
to which any discovery of his may lead. The
scientist, I think, is content to construct a perfectly harmonious picture on a mathematical
pattern, and he is quite satisfied to connect up the
various parts of it through mathematical formulas
without asking whether and how far these are a

proof that the law of causation functions in the external world.

MURPHY: Let me call your attention, Professor, to a phenomenon that happens sometimes down there on the lake when you are sailing your yacht. Of course it doesn't happen very often on the placid waters of Caputh, because you have flat lands all around and therefore no sudden windsqualls. But if you are sailing close to the wind on one of our northern lakes, you are always running the risk of heeling over rather suddenly under the onslaught of an unexpected air current. What I am coming to is, that I think the positivist might easily get in his shot here and hit you between wind and water. If you say that the scientist is content to secure mathematical logic in his mental construct, then you will quickly be quoted in support of the subjective idealism championed by modern scientists such as Sir Arthur Eddington.

EINSTEIN: But that would be ridiculous.

MURPHY: Of course it would be an unjustifiable conclusion; but you have already been widely quoted in the British Press as subscribing to the theory that the outer world is a derivative of consciousness. I have had to call this to the attention of a friend of mine in England, Mr. Joad, who has written an excellent book called *Philosophical Aspects of Science*. The book is a contradiction of the attitudes taken up by Sir Arthur Eddington and Sir James Jeans and your name is mentioned as corroborating their theories.

EINSTEIN: No physicist believes that. Otherwise he wouldn't be a physicist. Neither do the physicists you have mentioned. You must distinguish between what is a literary fashion and what is a scientific pronouncement. These men are genuine scientists and their literary formulations must not be taken as expressive of their scientific convictions. Why should anybody go to the trouble of gazing at the stars if he did not believe that the stars were really there? Here I am entirely at one with Planck. We cannot logically prove the existence of the external world, any more than you can logically prove that I am talking with you now or that I am here. But you know that I am here and no subjective idealist can persuade you to the contrary.

MURPHY: That point of course was fully elucidated long ago by the scholastics, and I cannot help thinking that much of the confusion in the nineteenth century and to-day would have been spared if the break with the philosophical tradition had not been so abysmal in the seventeenth century. The scholastics put the case for the modern physicist very clearly in describing mental images of external reality as existing *fundamentaliter in re, formaliter in mente.*

I forget how the discussion on this particular topic broke off. In the stenogram the next paragraph opens with PLANCK. There has recently been a great deal of discussion in the Press, I said to him, about what is called the bankruptcy of science. Is it that the general public here feel, somehow or other, that all the great scientific achievements of Germany seem to have been of no avail in securing the prestige of the nation abroad? Of course there is the larger background also of the general scepticism which is a universal feature of the world in our day. This attacks religion and art and literature as well as science.

PLANCK: The churches appear to be unable to supply that spiritual anchorage which so many people are seeking. And so the people turn in other directions. The difficulty which organized religion finds in appealing to the people nowadays is that its appeal necessarily demands the believing spirit, or what is generally called Faith. In an all-round state of scepticism this appeal receives only a poor response. Hence you have a number of prophets offering substitute wares.

MURPHY: Do you think that science in this particular might be a substitute for religion?

PLANCK: Not to a sceptical state of mind; for science demands also the believing spirit. Anybody who has been seriously engaged in scientific work of any kind realizes that over the entrance to the gates of the temple of science are written the words: *Ye must have faith*. It is a quality which the scientists cannot dispense with.

The man who handles a bulk of results obtained from an experimental process must have an imaginative picture of the law that he is pursuing. He must embody this in an imaginary hypothesis. The reasoning faculties alone will not help him forward a step, for no order can emerge from that chaos of elements unless there is the constructive quality of mind which builds up the order by a process of elimination and choice. Again and again the imaginary plan on which one attempts to build up that order breaks down and then we must try another. This imaginative vision and faith in the ultimate success are indispensable. The pure rationalist has no place here.

MURPHY: How far has this been verified in the lives of great scientists? Take the case of Kepler, whose 300th anniversary we were celebrating, you remember, that evening when Einstein gave his lecture at the Academy of Science. Wasn't there something about Kepler having made certain discoveries, not because he set out after them with his constructive imagination, but rather because he was concerned about the dimensions of wine barrels and was wondering which shapes would be the most economic containers?

PLANCK: These stories circulate in regard to nearly everybody whose name is before the public. As a matter of fact, Kepler is a magnificent example of what I have been saying. He was always hard up. He had to suffer disillusion after disillusion and even had to beg for the payment of the arrears of his salary by the Reichstag in Regensburg. He had to undergo the agony of having to defend his own mother against a public indictment of witchcraft. But one can realize, in studying his life, that what rendered him so energetic and tireless and productive was the profound faith he had in his own science, not the belief that he could eventually arrive at an arithmetical synthesis of his astronomical observations, but rather the profound faith in the existence of a definite plan behind the whole of creation. It was because he believed in that plan that his labour was felt by him to be worth while and also in this way, by never allowing his faith to lag, his work enlivened and enlightened his dreary life. Compare him with Tycho de Brahe. Brahe had the same material under his hands as Kepler, and even better opportunities, but he remained only a researcher, because he did not have the same faith in the existence of the eternal laws of creation. Brahe remained only a researcher; but Kepler was the creator of the new astronomy.

Another name that occurs to me in this connection is that of Julius Robert Mayer. His discoveries were hardly noticed, because in the middle of last century there was a great deal of scepticism, even among educated people, about the theories of natural philosophy. Mayer kept on and on, not because of what he had discovered and could prove, but because of what he believed. It was only in 1869 that the Society of German Physicists and Physicians, with Helmholtz at their head, recognized Mayer's work.

MURPHY: You have often said that the progress of science consists in the discovery of a new mystery the moment one thinks that something fundamental has been solved. The quantum theory has opened up this big problem of causation. And I really do not think that the matter can be answered very categorically. Of course it is easy enough to see that those who take up a definite stand and say that there is no such thing as causality are illogical, in the sense that you cannot prove any such statement either by experiment or by appeal to the direct dictates of consciousness and common sense in its defence. But, all the same, it seems to me that the burden is on the determinists at least to indicate the direction in which the old formulation of causality will have to be revised in order to meet the needs of modern science.

PLANCK: As to the first point, that about the discovery of new mysteries. This is undoubtedly true. Science cannot solve the ultimate mystery of nature. And that is because, in the last analysis, we ourselves are part of nature and therefore part of the mystery that we are trying to solve. Music and art are, to an extent, also attempts to solve or at least to express the mystery. But to my mind the more we progress with either the more we are brought into harmony with all nature itself. And that is one of the great services of science to the individual.

MURPHY: Goethe once said that the highest achievement to which the human mind can attain is an attitude of wonder before the elemental phenomena of nature.

PLANCK: Yes, we are always being brought face to face with the irrational. Else we couldn't have faith. And if we did not have faith but could solve every puzzle in life by an application of the human reason what an unbearable burden life would be. We should have no art and no music and no wonderment. And we should have no science; not only because science would thereby lose its chief attraction for its own followers— namely, the pursuit of the unknowable—but also because science would lose the cornerstone of its own structure, which is the direct perception by consciousness of the existence of external reality, As Einstein has said, you could not be a scientist if you did not know that the external world existed in reality; but that knowledge is not gained by any process of reasoning. It is a direct perception and therefore in its nature akin to what we call Faith. It is a metaphysical belief. Now that is something which the sceptic questions in regard to religion; but it is the same in regard to science. However, there is this to be said in favour of theoretical physics, that it is a very active science and does make an appeal to the lay imagination. In that way it may, to some extent, satisfy the metaphysical hunger which religion does not seem capable of satisfying nowadays. But this would be entirely by stimulating the religious reaction indirectly. Science as such can never really take the place of religion. This is explained in the penultimate chapter of the book.

MURPHY: And now for the second part of the question, that of the direction in which the traditional formulation of the causality principle may be revised. Einstein talks about the development of our faculties of perception as science goes on.

PLANCK: What exactly does he mean?

MURPHY: Perhaps I had better put it in my own way. Take for instance the modern phenomenon of speed. Fifty years ago the average tempo of locomotion was that of a trotting horse. Now it is even more than that of the railway train. If we strike a mean between the railway train and the motor-car and the aeroplane, we had better say sixty miles an hour instead of six miles an hour, as in the days of horse locomotion. You remember when bicycles first became popular. People were running down children and women on the roads day after day. Now you could not run down your grandmother with a bicycle. She'd be out of the way too quickly. You remember that when motors first careered along the roads the horses took fright. Now even the horses have developed their faculties to harmonize their perceptions with the idea of the new speed. There can be no doubt but that modern mankind has developed some faculty or other in regard to this new phenomenon of speed. Now I think Einstein's idea is that this sort of thing will go on developing, and that scientists will arise who will have a much keener perception than the scientists of to-day. They will, of course, also have more delicate instruments. But the point is that what we need to develop are the perceptive faculties themselves. It may be that a race of scientists trained in the laboratory will be able eventually to perceive the profound and manifold operation of causation in nature, just as the great musical genius perceives inner harmonies which the philistine cannot even dream of, and just as the music-lover can perceive keenly the beauty of a Beethoven symphonic structure, which the peasant could not appreciate at all, because he is accustomed only to his simple folk melodies. The development of the powers of perception therefore is one of the main tasks we have to meet. That seems to be Einstein's idea.

PLANCK: Of course it is clear. There is no doubt whatsoever that the stage at which theoretical physics has now arrived is beyond the average human faculties, even beyond the faculties of the great discoverers themselves. What, however, you must remember is that even if we progressed rapidly in the development of our powers of perception we could not finally unravel nature's mystery. We could see the operation of causation, perhaps, in the finer activities of the atoms, just as on the old basis of the causal formulation in classical mechanics we could perceive and make material images of all that was observed as occurring in nature.

Where the discrepancy comes in to-day is not between nature and the principle of causality, but rather between the picture which we have made of nature and the realities in nature itself. Our picture is not in perfect accord with our observational results; and, as I have pointed out over and over again, it is the advancing business of science to bring about a finer accord here. I am convinced that the bringing about of that accord must take place, not in the rejection of causality, but in a greater enlargement of the formula and a refinement of it, so as to meet modern discoveries.

INDEX